Introduction to Dislocations

Introduction to Dislocations

Fifth Edition

D. Hull and D. J. Bacon
*Department of Engineering,
Materials Science and Engineering,
University of Liverpool, UK*

Amsterdam • Boston • Heidelberg • London • New York • Oxford
Paris • San Diego • San Francisco • Singapore • Sydney • Tokyo
Butterworth-Heinemann is an imprint of Elsevier

Butterworth-Heinemann is an imprint of Elsevier
The Boulevard, Langford Lane, Kidlington, Oxford, OX5 1GB
30 Corporate Drive, Suite 400, Burlington, MA 01803, USA

First published 1965
Second edition 1975
Third edition 1984
Fourth edition 2001
Reprinted 2002, 2004, 2005, 2006, 2007, 2009 (twice)
Fifth edition 2011

Notices

Knowledge and best practice in this field are constantly changing. As new research and experience broaden our understanding, changes in research methods, professional practices, or medical treatment may become necessary.

Practitioners and researchers must always rely on their own experience and knowledge in evaluating and using any information, methods, compounds, or experiments described herein. In using such information or methods they should be mindful of their own safety and the safety of others, including parties for whom they have a professional responsibility.

To the fullest extent of the law, neither the Publisher nor the authors, contributors, or editors, assume any liability for any injury and/or damage to persons or property as a matter of products liability, negligence or otherwise, or from any use of operation of any methods, products, instructions, or ideas contained in the material herein.

British Library Cataloguing-in-Publication Data
A catalogue record for this book is available from the British Library

Library of Congress Cataloging-in-Publication Data
A catalog record for this book is available from the Library of Congress

ISBN: 978-0-08-096672-4

For information on all Butterworth-Heinemann publications
visit our website at www.elsevierdirect.com

Typeset by MPS Limited, a Macmillan Company, Chennai, India
www.macmillansolutions.com

Printed and bound in Great Britain
11 12 13 14 15   10 9 8 7 6 5 4 3 2 1

Contents

PREFACE..ix

CHAPTER 1 Defects in Crystals 1
  1.1 Crystalline Materials... 1
  1.2 Simple Crystal Structures 4
  1.3 Defects in Crystalline Materials............................. 8
      Point Defects 8
      Stacking Faults... 10
      Grain Boundaries 12
      Twin Boundaries.. 12
      Volume Defects 13
  1.4 Dislocations 13
      Geometry of Dislocations 15
      Burgers Vector and Burgers Circuit 16
  Further Reading 20

CHAPTER 2 Observation of Dislocations 21
  2.1 Introduction 21
  2.2 Electron Microscopy... 22
      General Principles... 22
      Dislocations 24
      Planar Defects 25
      Lattice Imaging 26
      Image Simulation 26
      Other Factors 28
  2.3 Other Experimental Methods................................. 28
      Surface Methods 28
      Decoration Methods.. 30
      X-ray Diffraction Topography 31
      Field Ion Microscopy and Atom
      Probe Tomography 32

2.4 Computer Simulation 34
      Atomic-Level Simulation 34
      Continuum-Level Simulation... 38
   Further Reading 41

CHAPTER 3   Movement of Dislocations 43
3.1 Concept of Slip 43
3.2 Dislocations and Slip...45
3.3 The Slip Plane..46
3.4 Cross Slip 48
3.5 Velocity of Dislocations..50
3.6 Climb..53
3.7 Experimental Observation of Climb....................................56
      Prismatic Dislocation Loops 56
      Helical Dislocations.. 57
3.8 Conservative Climb..59
3.9 Plastic Strain due to Dislocation Movement 59
   Further Reading 61

CHAPTER 4   Elastic Properties of Dislocations................................. 63
4.1 Introduction 63
4.2 Elements of Elasticity Theory 63
4.3 Stress Field of a Straight Dislocation..................................67
      Screw Dislocation... 67
      Edge Dislocation 69
4.4 Strain Energy of a Dislocation 71
4.5 Forces on Dislocations 73
4.6 Forces between Dislocations..75
4.7 Climb Forces..79
4.8 Image Forces 81
   Further Reading 82

CHAPTER 5   Dislocations in Face-centered Cubic Metals................. 85
5.1 Perfect Dislocations...85
5.2 Partial Dislocations – The Shockley Partial 85
5.3 Slip...87
5.4 Thompson's Tetrahedron 93
5.5 Frank Partial Dislocation...94
5.6 Dislocation Locks and Stair-Rod Partials..........................99
5.7 Stacking Fault Tetrahedra...103
   Further Reading 107

CHAPTER 6   Dislocations in Other Crystal Structures 109
6.1 Introduction 109
6.2 Dislocations in Hexagonal Close-packed Metals 109
    Burgers Vectors and Stacking Faults 109
    Basal and Non-basal Slip 112
    Vacancy and Interstitial Loops 115
6.3 Dislocations in Body-centered Cubic Metals 118
6.4 Dislocations in Ionic Crystals 124
6.5 Dislocations in Superlattices 126
6.6 Dislocations in Covalent Crystals 130
6.7 Dislocations in Layer Structures 133
6.8 Dislocations in Polymer Crystals................................ 134
Further Reading 135

CHAPTER 7   Jogs and the Intersection of Dislocations 137
7.1 Introduction 137
7.2 Intersection of Dislocations 137
7.3 Movement of Dislocations Containing
    Elementary Jogs.. 139
7.4 Superjogs 142
7.5 Jogs and Prismatic Loops.. 145
7.6 Intersections of Extended Dislocations and
    Extended Jogs 147
7.7 Attractive and Repulsive Junctions.............................. 149
7.8 Extended Stacking-fault Nodes.................................... 153
Further Reading 155

CHAPTER 8   Origin and Multiplication of Dislocations.................... 157
8.1 Introduction 157
8.2 Dislocations in Freshly Grown Crystals 158
8.3 Homogeneous Nucleation of Dislocations...................... 159
8.4 Nucleation of Dislocations at Stress Concentrators........ 160
8.5 Multiplication of Dislocations by
    Frank—Read Sources.. 163
8.6 Multiplication by Multiple Cross Glide 166
8.7 Multiplication by Climb 167
8.8 Grain Boundary Sources 169
Further Reading 169

CHAPTER 9   Dislocation Arrays and Crystal Boundaries 171
9.1 Plastic Deformation, Recovery and Recrystallization 171
9.2 Simple Dislocation Boundaries..................................... 175
9.3 General Low-angle Boundaries..................................... 177

9.4 Stress Field of Dislocation Arrays 181
9.5 Strain Energy of Dislocation Arrays..............................184
9.6 Dislocations and Steps in Interfaces..............................186
    Admissible Defects 186
    Identification of Interfacial Dislocations 189
    Dislocations in Epitaxial Interfaces 191
9.7 Movement of Boundaries...193
    Arrays of Crystal Dislocations.....................................193
    Glide of Interfacial Defects.......................................195
    Diffusion-assisted Motion of Interfacial Defects............199
    Martensitic Transformations....................................201
9.8 Dislocation Pile-ups...203
Further Reading 204

CHAPTER 10 Strength of Crystalline Solids.....................................205
10.1 Introduction 205
10.2 Temperature- and Strain-Rate-Dependence of
     the Flow Stress..207
10.3 The Peierls Stress and Lattice Resistance 210
    Dislocation Core Structure 210
    The Peierls Barrier 213
    Kink Mechanism 214
10.4 Interaction Between Point Defects and Dislocations ...216
10.5 Solute Atmospheres and Yield Phenomena 220
    Dislocation Locking......................................220
    Yield Drop 224
    Irradiation Hardening and Dislocation Channeling 227
10.6 The Flow Stress for Random Arrays of Obstacles 228
    Diffuse Forces...229
    Localized Forces...230
10.7 The Strength of Alloys...232
    Solutions, Precipitates and Aging 232
    Solution Strengthening.................................233
    Precipitate Strengthening 235
10.8 Work Hardening 237
10.9 Deformation of Polycrystals....................................243
10.10 Dislocations and Fracture......................................245
Further Reading...248

THE SI SYSTEM OF UNITS 251
INDEX 253

Preface

We are delighted to introduce the fifth edition of this book, which was first published in 1965 and has been in press continuously since then. In the Preface to the first edition it was noted that *"Although many aspects of the application of dislocation ideas are still in a state of flux a body of knowledge, based on theoretical analysis and experimental observation, has now been established which forms a basis for most of the ideas and theories which are proposed"*. Today the subject is a well-established discipline and is a powerful and essential aspect in understanding of the properties and behavior of crystalline solids.

Much of the basis of the subject, set out in relatively simple terms in the first edition, has been retained in later editions. In each subsequent edition new material has been included to broaden the detail and introduce some key aspects, which have emerged from research, without turning the book into a research monograph. In this latest edition we have taken advantage of recent research, using advanced computer modeling and very high resolution electron microscopy, to provide a better understanding of the arrangement of atoms close to the centre or core of dislocations. It is now clear that the atomic arrangement in the core has a fundamental effect on the behavior of dislocations and thereby on the properties of solids.

The development of this edition has benefited from a worldwide consultation, by the publishers, with leading authorities in the teaching and researching of dislocations. This consultation established that the fourth edition is still widely used in undergraduate and post-graduate university science and engineering courses and research schools. In preparing the fifth edition we have attempted to maintain a balance between a description of the basic properties of dislocations that are taught in introductory undergraduate courses and a more detailed treatment that is required for scientific research.

This distinction depends on the nature of courses and research although we have not attempted to make any delineation within the body of the text. Thus, for example, in Chapter 2, on the observation of dislocations, the simple methods, first developed in early studies, are set alongside the sophisticated approaches now available to specialists who have access to powerful analytical and experimental techniques, such as high resolution microscopy and codes that exploit the power of current computer systems to simulate dislocations at atomic and continuum scales.

A similar distinction has been made in the selection of further reading lists at the end of each chapter. We have included references to some of the influential papers from the early days of 'dislocations' as well as more recent papers that offer access to frontier research studies and provide an insight into the research now conducted in this field.

We are greatly indebted to friends and colleagues in many different countries who have allowed us access to their work and readily offered help and advice. We are also grateful to our publishers for encouraging us to prepare this new edition and agreeing to produce it in a completely new format.

Liverpool, August 2010   Derek Hull
                                                                             David Bacon

CHAPTER 1

Defects in Crystals

1.1 CRYSTALLINE MATERIALS

Dislocations are an important class of defect in crystalline solids and so an elementary understanding of crystallinity is required before dislocations can be introduced. Metals and many important classes of non-metallic solids are crystalline, i.e. the constituent atoms are arranged in a pattern that repeats itself periodically in three dimensions. The actual arrangement of the atoms is described by the *crystal structure*. The crystal structures of most pure metals are relatively simple: the three most common are the body-centered cubic, face-centered cubic and close-packed hexagonal, and are described in section 1.2. In contrast, the structures of alloys and non-metallic compounds are often complex.

The arrangement of atoms in a crystal can be described with respect to a three-dimensional net formed by three sets of straight, parallel lines as in Fig. 1.1(a). The lines divide space into equal sized parallelepipeds and the points at the intersection of the lines define a *space lattice*. Every point of a space lattice has identical surroundings. Each parallelepiped is called a *unit cell* and the crystal is constructed by stacking identical unit cells face to face in perfect alignment in three dimensions. By placing a *motif unit* of one or more atoms at every lattice site the regular structure of a perfect crystal is obtained.

The positions of the *planes*, *directions* and *point sites* in a lattice are described by reference to the unit cell and the three principal axes, x, y and z (Fig. 1.1 (b)). The cell dimensions $OA = a$, $OB = b$ and $OC = c$ are the lattice parameters, and these along with the angles $\angle BOC = \alpha$, $\angle COA = \beta$ and $\angle AOB = \gamma$ completely define the size and shape of the cell. For simplicity the discussion here will be restricted to cubic and hexagonal crystal structures. In cubic crystals $a = b = c$ and $\alpha = \beta = \gamma = 90°$, and the definition of planes and directions is straightforward. In hexagonal crystals it is convenient to use a different approach, and this is described in section 1.2.

Any *plane* $A'B'C'$ in Fig. 1.2 can be defined by the intercepts OA', OB' and OC' with the three principal axes. The usual notation (*Miller indices*) is to

1

Introduction to Dislocations.

Introduction to Dislocations.

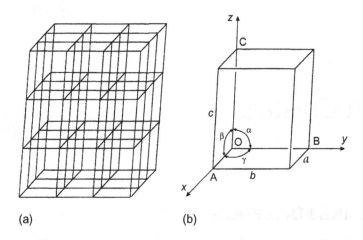

FIGURE 1.1

(a) A space lattice, (b) unit cell showing positions of principal axes.

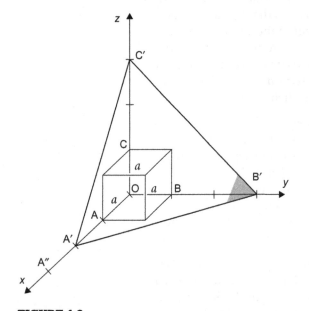

FIGURE 1.2

Cubic cell illustrating
method of describing the
orientation of planes.

take the reciprocals of the ratios of the intercepts to the corresponding unit cell dimensions. Thus $A'B'C'$ is represented by

$$\left(\frac{OA}{OA'}, \frac{OB}{OB'}, \frac{OC}{OC'}\right)$$

and the numbers are then reduced to the three smallest integers in these ratios.

Thus from Fig. 1.2 $OA' = 2a$, $OB' = 3a$, and $OC' = 3a$, the reciprocal intercepts are

$$\left(\frac{a}{2a}, \frac{a}{3a}, \frac{a}{3a}\right)$$

and so the Miller indices of the $A'B'C'$ plane are (322). Curved brackets are used for planes. A plane with intercepts OA, OB, and OC has Miller indices

$$\left(\frac{a}{a}, \frac{a}{a}, \frac{a}{a}\right)$$

or, more simply, (111). Similarly, a plane $DFBA$ in Fig. 1.3 is

$$\left(\frac{a}{a}, \frac{a}{a}, \frac{a}{\infty}\right)$$

or (110); a plane $DEGA$ is

$$\left(\frac{a}{a}, \frac{a}{\infty}, \frac{a}{\infty}\right)$$

or (100); and a plane $AB'C'$ in Fig. 1.2 is

$$\left(\frac{a}{a},\frac{a}{3a},\frac{a}{3a}\right)$$

or (311). In determining the indices of any plane it is most convenient to identify the plane of lattice points parallel to the plane which is closest to the origin O and intersects the principal axis close to the origin. Thus plane $A''B'C'$ in Fig. 1.2 is parallel to ABC and it is clear that the indices are (111). Using this approach it will be seen that the planes ABC, ABE, CEA and CEB in Fig. 1.3 are (111), $(11\bar{1})$, $(1\bar{1}1)$ and $(\bar{1}11)$ respectively. The minus sign above an index indicates that the plane cuts the axis on the negative side of the origin. In a cubic crystal structure, these planes constitute a group of the same crystallographic type and are described collectively by {111}.

FIGURE 1.3 Cubic cell illustrating the method of describing directions and point sites. *LM* is parallel to *OE*.

Any *direction LM* in Fig. 1.3 is described by the line parallel to *LM* through the origin O, in this case OE. The direction is given by the three smallest integers in the ratios of the lengths of the projections of OE resolved along the three principal axes, namely OA, OB and OC, to the corresponding lattice parameters of the unit cell. Thus, if the cubic unit cell is given by OA, OB and OC the direction LM is

$$\left[\frac{OA}{OA},\frac{OB}{OB},\frac{OC}{OC}\right]$$

or

$$\left[\frac{a}{a},\frac{a}{a},\frac{a}{a}\right]$$

or [111]. Square brackets are used for directions. The directions CG, AF, DB and EO are $[11\bar{1}]$, $[\bar{1}11]$, $[\bar{1}1\bar{1}]$ and $[\bar{1}\bar{1}\bar{1}]$ respectively and are a group of directions of the same crystallographic type described collectively by $\langle 111\rangle$. Similarly, direction CE is

$$\left[\frac{a}{a},\frac{a}{a},\frac{O}{a}\right]$$

or [110]; direction AG is

$$\left[\frac{O}{a},\frac{a}{a},\frac{O}{a}\right]$$

or [010]; and direction GH is

$$\left[\frac{-a/2}{a},\frac{-a/2}{a},\frac{a}{a}\right]$$

or $[\overline{11}2]$. The rule that brackets [ ] and ( ) imply specific directions and planes respectively, and that $\langle \rangle$ and { } refer respectively to directions and planes of the same type, will be used throughout this text.

In cubic crystals the Miller indices of a plane are the same as the indices of the direction normal to that plane. Thus in Fig. 1.3 the indices of the plane *EFBG* are (010) and the indices of the direction *AG* which is normal to *EFBG* are [010]. Similarly, direction *OE* [111] is normal to plane *CBA* (111).

The coordinates of any *point* in a crystal relative to a chosen origin site are described by the fractional displacements of the point along the three principal axes divided by the corresponding lattice parameters of the unit cell. The center of the cell in Fig. 1.3 is $\frac{1}{2}, \frac{1}{2}, \frac{1}{2}$ relative to the origin O; and the points F, E, H and I are 0, 1, 1; 1, 1, 1; $\frac{1}{2}, \frac{1}{2}, 1$; and 1, $\frac{1}{2}$, 1 respectively.

1.2 SIMPLE CRYSTAL STRUCTURES

In this section the atoms are considered as hard spheres which vary in size from element to element. From the hard sphere model the parameters of the unit cell can be described directly in terms of the radius of the atomic sphere, r. In the diagrams illustrating the crystal structures the atoms are shown as small circles in the three-dimensional drawings and as large circles representing the full hard sphere sizes in the two-dimensional diagrams. It will be shown that crystal structures can be described as a stack of lattice planes in which the arrangement of lattice sites within each layer is identical. To see this clearly in two-dimensional figures, the atoms in one layer represented by the plane of the paper are shown as full circles, whereas those in layers above and below the first are shown as small shaded circles. The order or sequence of the atom layers in the stack, i.e. the *stacking sequence*, is described by labeling one layer as an *A* layer and all other layers with atoms in identical positions above the first as *A* layers also. Layers of atoms in other positions in the stack are referred to as *B*, *C*, *D* layers, etc.

In the *simple cubic structure* with one atom at each lattice site, illustrated in Fig. 1.4, the atoms are situated at the corners of the unit cell. (Note that no real crystals have such a simple atomic arrangement.) Figures 1.4(b) and (c) show the arrangements of atoms in the (100) and

FIGURE 1.4

Simple cubic structure: (a) unit cell, (b) arrangement of atoms in (100) layers, (c) arrangement of atoms in (110) layers.

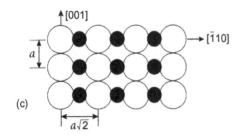

(110) planes respectively. The atoms touch along ⟨001⟩ directions and therefore the lattice parameter a is twice the atomic radius r $(a = 2r)$. The atoms in adjacent (100) planes are in identical atomic sites when projected along the direction normal to this plane, so that the stacking sequence of (100) planes is AAA... The atoms in adjacent (110) planes are displaced $\frac{1}{2} a\sqrt{2}$ along $[\bar{1}10]$ relative to each other and the spacing of atoms along [110] is $a\sqrt{2}$. It follows that alternate planes have atoms in the same atomic sites relative to the direction normal to (110) and the stacking sequence of (110) planes is $ABABAB$... The spacing between successive (110) planes is $\frac{1}{2} a\sqrt{2}$.

In the *body-centered cubic structure* (bcc), which is exhibited by many metals and is shown in Fig. 1.5, the atoms are situated at the corners of the unit cell and at the centre site $\frac{1}{2}, \frac{1}{2}, \frac{1}{2}$. The atoms touch along a ⟨111⟩ direction and this is referred to as the *close-packed direction*. The lattice parameter $a = 4r/\sqrt{3}$ and the spacing of atoms along ⟨110⟩ directions is $a\sqrt{2}$. The stacking sequence of {100} and {110} planes is $ABABAB$... (Fig. 1.5(b)). There is particular interest in the stacking of {112} type planes (see sections 6.3 and 9.7). Figure 1.6 shows two body-centered cubic cells and the positions of a set of $(1\bar{1}2)$ planes. From the diagrams it is seen that the stacking sequence of these planes is $ABCDEFAB$... , and the spacing between the planes is $a/\sqrt{6}$.

In the *face-centered cubic structure* (fcc), which is also common among the metals and is shown in Fig. 1.7, the atoms are situated at the corners of the unit cell and at the centers of all the cube faces in sites of the type 0, $\frac{1}{2}, \frac{1}{2}$. The atoms touch along the ⟨011⟩ close-packed directions. The lattice parameter $a = 4r/\sqrt{2}$. The stacking sequence of {100} and {110} planes is

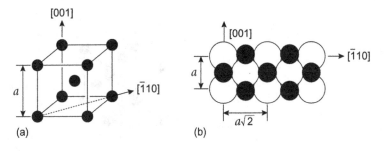

FIGURE 1.5

Body-centered cubic structures: (a) unit cell, (b) arrangement of atoms in (110) layers.

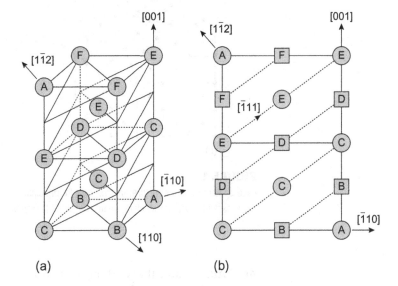

FIGURE 1.6

Stacking sequence of {112} planes in a body-centered cubic crystal. (a) Two unit cells showing positions of atoms in $(1\bar{1}2)$ planes. (b) Traces of the $(1\bar{1}2)$ planes on a (110) projection: atom sites marked by circles lie in the plane of the diagram; those marked by squares lie $a/\sqrt{2}$ above and below.

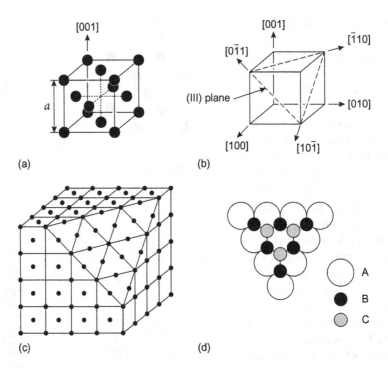

FIGURE 1.7

Face-centered cubic structure: (a) unit cell, (b) principal directions, (c) arrangement of atoms in a (111) close-packed plane, (d) stacking sequence of {111} planes.

ABABAB..., and the stacking sequence of {111} planes is *ABCABC...* The latter is of considerable importance (see Chapter 5) and is illustrated in Figs 1.7(c) and (d). The atoms in the {111} planes are in the most close-packed arrangement possible for spheres and contain three ⟨110⟩ close-packed directions 60° apart, as in Fig. 1.7(b).

The *close-packed hexagonal structure* (cph or hcp) is also common in metals. It is more complex than the cubic structures but can be described very simply with reference to the stacking sequence. The unit cell with lattice parameters a, a, c is shown in Fig. 1.8(a), together with the hexagonal cell constructed from three unit cells. There are two atoms per lattice site, i.e. at 0, 0, 0 and $\frac{2}{3}, \frac{1}{3}, \frac{1}{2}$ with respect to the axes a_1, a_2, c. The atomic planes perpendicular to the c axis are close-packed, as in the fcc case, but the stacking sequence is now *ABABAB...*, as shown in Fig. 1.8(b).

For a hard sphere model the ratio of the length of the c and a axes (axial ratio) of the hexagonal structure is 1.633. In practice, the axial ratio varies

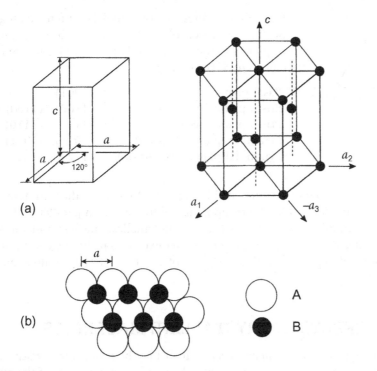

FIGURE 1.8
Close-packed hexagonal structure: (a) the unit cell of the lattice and the hexagonal cell showing the arrangement of atoms, (b) *ABAB...* stacking sequence of the atomic planes perpendicular to the *c* axis.

between 1.57 and 1.89 in close-packed hexagonal metals. The variations arise because the hard sphere model gives only an approximate value of the interatomic distances and requires modification depending on the electronic structure of the atoms.

If Miller indices of three numbers based on axes a_1, a_2, c are used to define planes and directions in the hexagonal structure, it is found that crystallographically equivalent sets can have combinations of different numbers. For example, the three close-packed directions in the basal plane (001) are [100], [010] and [110]. Indexing in hexagonal crystals is therefore usually based on *Miller-Bravais* indices, which are referred to the four axes a_1, a_2, a_3 and c indicated in Fig. 1.8(a). When the reciprocal intercepts of a plane on all four axes are found and reduced to the smallest integers, the indices are of the type (h, k, i, l), and the first three indices are related by

$$i = -(h + k) \tag{1.1}$$

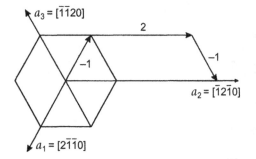

FIGURE 1.9

Determination of direction indices in the basal plane of an hexagonal crystal. The translations giving rise to $[\bar{1}2\bar{1}0]$ are shown explicitly.

Equivalent planes are obtained by interchanging the position and sign of the first three indices. A number of planes in the hexagonal lattice have been given specific names. For example:

Basal plane	(0001)
Prism plane : first order	$(1\bar{1}00)$ $(\bar{1}100)$, etc.
Prism plane : second order	$(11\bar{2}0)$ $(\bar{2}110)$, etc.
Pyramidal plane : first order	$(10\bar{1}1)$ $(\bar{1}011)$, etc.
Pyramidal plane : second order	$(11\bar{2}2)$ $(\bar{1}\,\bar{1}22)$, etc.

Some of these planes are indicated in Fig. 6.1. Direction indices in hexagonal structures are defined by the components of the direction parallel to the four axes. The numbers must be reduced to the smallest integers and the third index is the negative of the sum of the first two. To satisfy this condition the directions along axes a_1, a_2 and a_3 are of the type $\langle\bar{1}2\bar{1}0\rangle$ as illustrated in Fig. 1.9.

1.3 DEFECTS IN CRYSTALLINE MATERIALS

All real crystals contain *imperfections* which may be *point, line, surface* or *volume* defects, and which disturb locally the regular arrangement of the atoms. Their presence can significantly modify the properties of crystalline solids, and although this text is primarily concerned with the line defects called *dislocations*, it will be seen that the behavior and effects of all these imperfections are intimately related.

Point Defects

All the atoms in a perfect crystal are at specific atomic sites (ignoring thermal vibrations). In a pure metal two types of point defect are possible, namely a *vacant atomic site* or *vacancy*, and a *self-interstitial atom*. These *intrinsic* defects are illustrated for a simple cubic structure in Fig. 1.10. The vacancy has been formed by the removal of an atom from an atomic site (labeled v) and the interstitial by the introduction of an atom into a non-lattice site at a $\frac{1}{2}, \frac{1}{2}, 0$ position (labeled i). It is known that vacancies and interstitials can be produced in materials by plastic deformation and high-energy particle irradiation. The latter process is particularly important in materials in nuclear reactor installations. Furthermore, intrinsic point defects are introduced into crystals simply by virtue of temperature, for at all temperatures above 0 K there is a thermodynamically stable concentration. The change in Helmholtz free energy ΔF associated with the introduction of n vacancies or self-interstitials in the crystal is

$$\Delta F = nE_f - T\Delta S \qquad (1.2)$$

where E_f is the energy of formation of one defect and ΔS is the change in the entropy of the crystal. nE_f represents a considerable positive energy, but this is offset by an increase in the configurational entropy due to the presence of the defects. The equilibrium concentration of defects, given by the ratio of the number of defects to the number of atomic sites, corresponding to the condition of minimum free energy is approximately

$$c_0 = \exp\left(-\frac{E_f}{kT}\right) \qquad (1.3)$$

where k is Boltzmann's constant and T is the temperature (in deg K).

(a)   (b)

FIGURE 1.10

(a) Vacancy, (b) self-interstitial atom in an (001) plane of a simple cubic crystal.

For the vacancy, the formation energy, E_f^v, is that required to remove one atom from its lattice site to a place on the surface of the crystal. Experimental values fall in the range ~1−3 eV, i.e. 0.16−0.48 aJ. They scale with the melting temperature, T_m, and an approximate rule is $E_f^v \simeq 8kT_m$. Thus for copper, for which $T_m = 1356$ K and $E_f^v = 1.3$eV, the fraction of atom sites vacant at 1300 K is ~10^{-5} and at 300 K is ~10^{-22}. The formation energy of the self-interstitial is the energy required to remove one atom from the surface and insert it into an interstitial site, and E_f^i is typically two to four times E_f^v. Consequently, the concentration given by equation (1.3) is many orders of magnitude smaller for interstitials, and so in metals in thermal equilibrium the concentration of interstitials may be neglected in comparison with that of vacancies. In non-metals, ionic charge and valence effects may modify this conclusion.

The rate at which a point defect moves from site to site in the crystal is proportional to $\exp(-E_m/kT)$, where E_m is the defect *migration energy* and is typically ~0.1−1.0 eV. The rate decreases exponentially with decreasing temperature and consequently in many metals it is possible to retain a high vacancy concentration at room temperature by rapidly quenching from a high equilibrating temperature.

Impurity atoms in a crystal can be considered as *extrinsic* point defects and they play an important role in the physical and mechanical properties of all materials. Impurity atoms can take up two different types of site, as illustrated in Fig. 1.11: (a) *substitutional*, in which an atom of the parent lattice is replaced by the impurity atom, and (b) *interstitial*, in which the impurity atom is at a non-lattice site similar to the self-interstitial atoms referred to above.

All the point defects mentioned produce a local distortion in the otherwise perfect crystal. The amount of distortion and hence the amount of additional

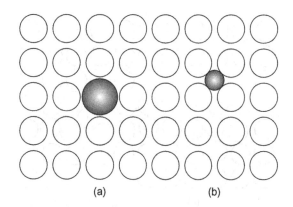

FIGURE 1.11

(a) Substitutional impurity atom, (b) interstitial impurity atom.

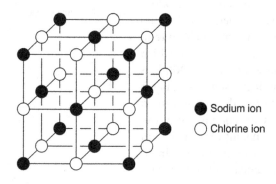

Sodium ion

Chlorine ion

FIGURE 1.12

Sodium chloride structure which consists of two interpenetrating face-centered cubic lattices of the two types of atom, with the corner of one located at the point $\frac{1}{2}$, 0, 0 of the other.

energy in the crystal due to the defects depends on the amount of 'space' between the atoms and the 'size' of the atoms introduced.

The interstice sites between atoms generally have volumes of less than one atomic volume, and the interstitial atoms therefore tend to produce large distortions among the surrounding atoms. This accounts for the relatively large values of E_f^i referred to above for the self-interstitial, and can result in crystal volume increases as large as several atomic volumes per interstitial atom.

Additional effects are important when the removal or addition of atoms changes the local electric charge in the crystal. This is relatively unimportant in crystals with metallic binding, but can be demonstrated particularly well in crystals in which the binding is ionic. The structure of sodium chloride is shown in Fig. 1.12. Each negatively charged chlorine ion is surrounded by six nearest neighbors of positively charged sodium ions and vice versa. The removal of a sodium or a chlorine ion produces a local negative or positive charge as well as a vacant lattice site. These are called *cation* and *anion* vacancies respectively. To conserve an overall neutral charge the vacancies must occur either (a) in pairs of opposite sign, forming divacancies known as *Schottky defects*, or (b) in association with interstitials of the same ion, *Frenkel defects*.

Stacking Faults

In section 1.2 it was emphasized that perfect crystals can be described as a stack of atom layers arranged in a regular sequence. For the simple metallic structures discussed in section 1.2, the atomic layers are identical. A *stacking fault* is a *planar defect* and, as its name implies, it is a local region in the crystal where the regular stacking sequence is interrupted. Stacking faults are not expected in planes with *ABABAB...* sequences in body-centered or face-centered cubic metals because there is no alternative site for an *A* layer resting on a *B* layer. However, for *ABCABC...* or *ABABAB...* stacking of the close-packed planes in close-packed structures there are two possible positions of one layer resting on another (Fig. 1.7). According to the hard sphere

FIGURE 1.13

Stacking faults in the face-centered cubic structure. The normal stacking sequence of (111) planes is denoted by *ABCA*... Planes in normal relation to one another are separated by $\triangle$, those with a stacking error by $\triangledown$: (a) intrinsic stacking fault, (b) extrinsic stacking fault.

model, a close-packed layer of atoms resting on an A layer can rest equally well in either a B or a C position and geometrically there is no reason for the selection of a particular position. In a face-centered cubic lattice two types of stacking fault are possible, referred to as *intrinsic* and *extrinsic*. These are best described by considering the change in sequence resulting from the removal or introduction of an extra layer. In Fig. 1.13(a) part of a C layer has been removed; this results in a break in the stacking sequence. This is an *intrinsic fault* and it can be seen that the stacking sequences above and below the fault plane are continuous right up to the fault itself. In Fig. 1.13(b) an extra A layer has been introduced between a B and a C layer. There are two breaks in the stacking sequence and it is referred to as an *extrinsic fault*. The extra layer does not belong to the continuing patterns of the lattice either above or below the fault.

The presence of stacking faults can play an important role in the plasticity of crystals. It should be noted, for example, that the intrinsic fault in the face-centered cubic structure can be produced by a sliding process. If, say, an A layer is slid into a B position and all the layers above are moved in the same way, i.e. B to C, C to A, A to B, etc., then the sequence ABCABCA... becomes *ABCBCAB*..., which is identical to that discussed above. This aspect will be discussed in more detail in Chapter 5.

Stacking faults have been reported in many crystal structures. (Those which occur in the close-packed hexagonal metals are described in Chapter 6.) They destroy the perfection of the host crystal, and the associated energy per unit area of fault is known as the *stacking-fault energy*. Typical values lie in the range $1-1000$ mJ/m^2. In the faults described above, the first- and second-nearest-neighbor atomic bonds in the close-packed structure are preserved, so that only bonds to more-distant neighbors and electronic effects contribute to the energy. In other structures, covalent bonding and ionic effects can be important.

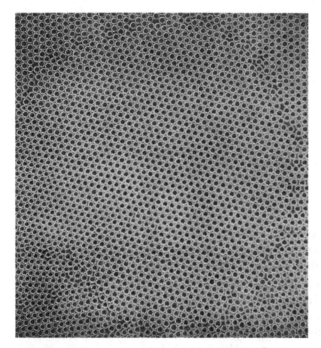

FIGURE 1.14

Crystal grains simulated by a bubble raft. *(From* Scientific American, *Sept. 1967.)*

FIGURE 1.15

Arrangement of atoms in a twin related structure; $x-y$ is the trace of the twin composition plane.

Grain Boundaries

Crystalline solids usually consist of a large number of randomly oriented *grains* separated by grain boundaries. Each grain is a single crystal and contains the defects already described. When the misorientation between the grains is large, the atomic arrangement at the boundary is complicated and varies significantly with the angle of misorientation. An easy way to visualize the atomic arrangement is to use bubble models (Fig. 1.14) in which a two-dimensional raft of equal sized bubbles floats on the surface of a liquid. Figure 1.14 shows a grain or 'crystal' surrounded by grains of different orientation. A notable feature of the boundary structure is that the region of disorder is very narrow, being limited to one or two 'atoms' on each side of the boundary. For certain misorientation relationships between grains, the structure of the boundary can be described as an array of dislocations, as discussed in Chapter 9.

Twin Boundaries

Deformation twinning is a process in which a region of a crystal undergoes a homogeneous shear that produces the original crystal structure in a new orientation. In the simplest cases, this results in the atoms of the original crystal ('parent') and those of the product crystal ('twin') being mirror images of each other by reflection in a *composition plane*, as illustrated in Fig. 1.15. The open circles represent the positions of the atoms before twinning and the black circles the positions after twinning. The atoms above $x-y$ are mirror images of the atoms below and therefore $x-y$ represents the trace of the twin composition plane in the plane of the paper. The homogeneous shear of the lattice parallel to the composition plane is denoted by arrows. Deformation twinning can be induced by plastic deformation and is particularly important in body-centered cubic and close-packed hexagonal metals and many non-metallic crystals (Chapter 6). When a growing twin meets a flat surface it produces a well-defined tilt, and this can readily be detected in an optical microscope. Figure 1.16 shows the tilts produced by deformation twins in a 3.25 per cent silicon iron crystal deformed at 20 K. Although the twinning process differs from slip (Chapter 3), in which there is no rotation of the lattice, the sequential shear of atoms in planes parallel to the composition plane (Fig. 1.15) occurs by the movement of twinning dislocations (Chapter 9).

Volume Defects

Crystal defects such as precipitates, voids and bubbles can occur under certain circumstances and have important effects on the properties of crystalline solids. As an example, it will be seen in Chapter 10 how the interaction of dislocations with precipitates has played a vital role in the development of high-strength alloys.

FIGURE 1.16

Deformation twins in 3.25 per cent silicon iron. The surface at the twins is tilted so light is reflected away from the objective lens of the microscope.

1.4 DISLOCATIONS

Although there are many techniques now available for the direct observation of dislocations (Chapter 2), the existence of these line defects was deduced by inference in the early stages of dislocation study (1934 to the early 1950s). Strong evidence arose from attempts to reconcile theoretical and experimental values of the applied shear stress required to *plastically deform* a single crystal. As explained in section 3.1, this deformation occurs

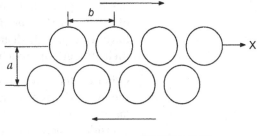

FIGURE 1.17

Representation of atom positions used to estimate the theoretical critical shear stress for slip.

by atomic planes sliding over each other. In a perfect crystal, i.e. in the absence of dislocations, the sliding of one plane past an adjacent plane would have to be a rigid co-operative movement of all the atoms from one position of perfect registry to another. The shear stress required for this process was first calculated by Frenkel in 1926. The situation is illustrated in Fig. 1.17. It is assumed that there is a periodic shearing force required to move the top row of atoms across the bottom row which is given by the sinusoidal relation:

$$\tau = \frac{Gb}{2\pi a}\sin\frac{2\pi x}{b} \qquad (1.4)$$

where τ is the applied shear stress, G is the shear modulus, b the spacing between atoms in the direction of the shear stress, a the spacing of the rows of atoms and x is the shear translation of the two rows away from the low-energy position of stable equilibrium ($x = 0$).

The right-hand side of equation (1.4) is periodic in b and reduces to Hooke's law (section 4.2) for small strains x/a, i.e. in the small-strain limit $\sin(2\pi x/b) \simeq (2\pi x/b)$ and so $\tau = Gx/a$. The maximum value of τ is then the *theoretical critical shear stress* and is

$$\tau_{th} = \frac{b}{a}\frac{G}{2\pi} \tag{1.5}$$

Since $b \approx a$, the theoretical shear strength is a sizeable fraction of the shear modulus. Using more realistic expressions for the force as a function of shear displacement, values of $\tau_{th} \approx G/30$ are obtained. Although these are approximate calculations, they show that τ_{th} is many orders of magnitude greater than the observed values (10^{-4} to 10^{-8} G) of the resolved shear stress for slip measured in real, well-annealed crystals. This striking difference between prediction and experiment was accounted for independently by Orowan, Polanyi and Taylor in 1934 by the presence of dislocations. Since then, it has been possible to produce crystals in the form of fibers with a diameter of a few µm ('*microwhiskers*') or even less ('*nanowires*' or '*nanopillars*'), which have a very high degree of perfection. When entirely free of dislocations, their strength approaches the theoretical strength.

Other evidence which contributed appreciably to the universal acceptance of the existence of dislocations in crystals, was the reconciliation of the classical *theory of crystal growth* with the experimental observations of growth rates. Consider a perfect crystal having irregular facets growing in a supersaturated vapor. At a low degree of supersaturation, growth occurs by the deposition of atoms on the irregular or imperfect regions of the crystal. The preferential deposition in imperfect regions results in the formation of more perfect faces consisting of close-packed arrays of atoms. Further growth then requires the nucleation of a new layer of atoms on a smooth face. This is a much more difficult process, and nucleation theory predicts that for growth to occur at the observed rates a degree of supersaturation of approximately 50 per cent would be required. This is contrary to many experimental observations which show that growth occurs readily at a supersaturation of only 1 per cent. The difficulty was resolved when it was demonstrated that the presence of dislocations in the crystal during growth could result in the formation of steps on the crystal faces which are not removed by preferential deposition, as in a perfect crystal. As explained in section 2.3 (Fig. 2.10(c)), these steps provide sites for deposition and thus eliminate the difficult nucleation process.

Geometry of Dislocations

The role of dislocations in plastic deformation is explained in Chapter 3. At this stage it will be sufficient to describe the basic geometry of an *edge* and a *screw* dislocation line and introduce the appropriate definitions and terminology.

Figure 1.18(a) represents an elementary descriptive model of the atomic arrangement and bonding in a simple cubic structure. For convenience it is

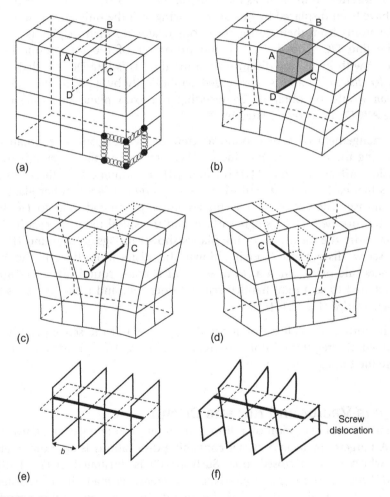

(a)
(b)
(c)
(d)
(e)
(f)

Screw dislocation

FIGURE 1.18

(a) Model of a simple cubic lattice; the atoms are represented by filled circles, and the bonds between atoms by springs, only a few of which are shown; (b) positive edge dislocation *DC* formed by inserting an extra half-plane of atoms in *ABCD*; (c) left-handed screw dislocation *DC* formed by displacing the faces *ABCD* relative to each other in direction *AB*; (d) right-handed screw dislocation *DC*; (e) atomic planes, spacing *b*, in a perfect crystal; (f) planes distorted by a right-handed screw dislocation.

assumed that the bonds can be represented by flexible springs between adjacent atoms. It must be emphasized that bonding in real solids is complex and, in fact, the nature of the bonding determines the fine detail of the arrangement of the atoms around the dislocation. The arrangement of atoms around an edge dislocation can be simulated by the following sequence of operations. Suppose that all the bonds across the surface *ABCD* are broken and the faces of the crystal are separated so that an *extra half-plane of atoms* can be inserted in the slot, as illustrated in Fig. 1.18(b). The faces of the slot will have been displaced by one atom spacing, but the only large disturbance of the atoms from their normal positions relative to their neighbors is close to the line *DC*. The deflection and distortion of the interatomic bonds decrease with increasing distance from the line. This line *DC* is called a *positive edge dislocation* and is represented symbolically by ⊥. A *negative edge dislocation* would be obtained by inserting the extra plane of atoms below plane *ABCD* and is represented by ⊤.

The arrangement of atoms round a screw dislocation can be simulated by displacing the crystal on one side of *ABCD* relative to the other side in the direction *AB* as in Figure 1.18(c) and (d). Examination of these models shows that each can be described as a *single surface helicoid*, rather like a spiral staircase. The set of parallel planes initially perpendicular to *DC* have been transformed into a single surface, and the spiral nature is clearly demonstrated by the distortion of the planes shown in Figs 1.18(e) and (f). *DC* is a screw dislocation. Looking down the dislocation line, if the helix advances one plane when a *clockwise circuit* is made round it (Fig. 1.18(d)), it is referred to as a right-handed screw dislocation, and if the reverse is true it is *left-handed* (Fig. 1.18(c)).

It is important to realize that for both the edge and the screw dislocations described, the registry of atoms across the interface *ABCD* is identical to that before the bonds were broken.

Burgers Vector and Burgers Circuit

The most useful definition of a dislocation is given in terms of the *Burgers circuit*. A Burgers circuit in a crystal containing dislocations is an atom-to-atom path which forms a closed loop. Such a path is illustrated in Fig. 1.19(a), i.e. *MNOPQ*. If the same atom-to-atom sequence is made in a dislocation-free crystal, Figure 1.19(b), and the circuit does not close, then the first circuit, Fig. 1.19(a), must enclose one or more dislocations. The vector required to complete the circuit is called the *Burgers vector*. It is essential that the circuit in the real crystal passes entirely through 'good' parts of the crystal. For simplicity consider the Burgers circuit to enclose one dislocation as in Fig. 1.19(a). The sequence of atom-to-atom movements in the perfect

FIGURE 1.19

(a) Burgers circuit round an edge dislocation with positive line sense into the paper (see text); (b) the same circuit in a perfect crystal; the closure failure is the Burgers vector.

FIGURE 1.20

(a) Burgers circuit round a left-handed screw dislocation with positive line sense in the direction shown; (b) the same circuit in a perfect crystal; the closure failure is the Burgers vector.

crystal is the same as for the circuit *MNOPQ* in Fig. 1.19(a). The closure failure *QM* is the Burgers vector and is at *right angles* to the dislocation line (cf. Fig. 1.18(b)). When the Burgers circuit is drawn round a screw disloca-tion (Fig. 1.20), again with a closed circuit in the crystal containing the dis-location, the Burgers vector *QM* is *parallel* to the dislocation line. This leads to two important rules:

(a) *The Burgers vector of an edge dislocation is normal to the line of the dislocation.*

(b) *The Burgers vector of a screw dislocation is parallel to the line of the dislocation.*

In the most general case (Chapter 3) the dislocation line lies at an arbitrary angle to its Burgers vector and the dislocation line has a *mixed* edge and screw character. However, the Burgers vector of a single dislocation has fixed length and direction, and is independent of the position and orientation of the dislocation line.

Burgers circuits taken around other defects, such as vacancies and interstitials, do not lead to closure failures. Two rules are implied by the Burgers circuit construction used above. First, when looking along the dislocation line, which defines the positive *line sense* or direction of the dislocation, the circuit is taken in a clockwise fashion (Figs 1.19(a), 1.20(a)). Second, the Burgers vector is taken to run from the finish to the start point of the reference circuit in the perfect crystal. This defines the right-hand/finish-start (RH/FS) convention. It is readily shown by use of sketches similar to those of Figs 1.19 and 1.20, that *reversing the line sense reverses the direction of the Burgers vector* for a given dislocation. Furthermore, *dislocations with the same line sense but opposite Burgers vectors* (or alternatively with opposite line senses and the same Burgers vector) *are physical opposites*, in that if one is a positive edge, the other is a negative edge, and if one is a right-handed screw, the other is left-handed. Dislocations which are physical opposites of each other annihilate and restore perfect crystal if brought together.

The Burgers vectors defined in the simple cubic crystals of Figs 1.19 and 1.20 are the shortest *lattice translation vectors* which join two points in the lattice. A dislocation whose Burgers vector is a lattice translation vector is known as a *perfect* or *unit dislocation*. The Burgers vector **b** is conveniently described using the indices defined in section 1.1. For example, the lattice vector from the origin to the center of a body-centered cubic cell is defined both in magnitude and direction by displacements of $a/2$ in the x-direction, $a/2$ in the y-direction and $a/2$ in the z-direction, and the notation used is $b = \frac{1}{2}[111]$. The magnitude (or length) b of the vector is

$$b = \sqrt{\left(\frac{a^2}{4} + \frac{a^2}{4} + \frac{a^2}{4}\right)} = \frac{a\sqrt{3}}{2} \tag{1.6}$$

Similarly, if **b** is the shortest lattice translation vector in the face-centered cubic structure, i.e. $b = \frac{1}{2}\langle 110 \rangle$, then $b = a/\sqrt{2}$.

Dislocation lines can end at the surface of a crystal and at grain boundaries, but never inside a crystal. Thus, *dislocations must either form closed loops or branch into other dislocations.* When three or more dislocations meet at a point, or *node*, it is a necessary condition that the Burgers vector is conserved, i.e. vector total in equals vector total out. Consider the dislocation $\mathbf{b}_1$ (Fig. 1.21) which branches into two dislocations with Burgers vectors $\mathbf{b}_2$ and $\mathbf{b}_3$. A Burgers circuit has been drawn round each according to the line senses indicated. The large circuit on the right-hand side of the diagram encloses two dislocations, but since it passes through the same good material as the $\mathbf{b}_1$ circuit on the left-hand side the Burgers vector must be the same, i.e. $\mathbf{b}_1$. It follows from the diagram that

$$\mathbf{b}_1 = \mathbf{b}_2 + \mathbf{b}_3 \tag{1.7}$$

It is more usual to define the Burgers circuits by making a clockwise circuit around each dislocation line looking outward from the nodal point. This reverses the line sense (and hence $\mathbf{b}_1$) on the left-hand side, and then equation (1.7) becomes

$$\mathbf{b}_1 + \mathbf{b}_2 + \mathbf{b}_3 = 0 \tag{1.8}$$

or, more generally, for n dislocation branches

$$\sum_1^n \mathbf{b}_i = 0 \tag{1.9}$$

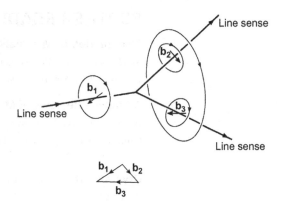

FIGURE 1.21
Three dislocations forming a node.

The *dislocation density* ρ is defined as the total length of dislocation line per unit volume of crystal, normally quoted in units of cm^{-2} or m^{-2} ($1\ \text{m}^{-2} \equiv 10^4\ \text{cm}^{-2}$). Thus for a volume V containing line length l, $\rho = l/V$. All crystals, apart from some nanowires, contain dislocations and in well-annealed crystals the dislocations are arranged in a rather ill-defined network, the *Frank net*, as illustrated schematically in Fig. 1.22. In well-annealed metal crystals ρ is usually between 10^{10} and $10^{12}\ \text{m}^{-2}$, but it increases rapidly with plastic deformation, and a typical value for a heavily cold-rolled metal is about 10^{14} to $10^{15}\ \text{m}^{-2}$. The arrangement of the dislocations depends on the conditions of loading and some actual examples are presented elsewhere in the book. ρ is usually lower in non-metallic crystals than in metal crystals, and values down to $10^5\ \text{m}^{-2}$ can be obtained in carefully grown semiconductor crystals.

An alternative definition of ρ, which is sometimes more convenient to use, is the number of dislocations intersecting unit area of a planar surface within the crystal. If all the dislocations are parallel, the two density values are the same, but for a completely random arrangement the volume density is twice the surface density. The surface density provides a convenient way of estimating the average distance between dislocations in a network of density ρ. With ρ intersections per unit area, the area per intersection is $1/\rho$ and so the distance between neighboring dislocations is of the order of $1/\sqrt{\rho}$, i.e. 10^{-3}, 10^{-5} and 10^{-7} m for ρ equal to 10^6, 10^{10} and $10^{14}\ \text{m}^{-2}$, respectively.

FIGURE 1.22
Diagrammatic illustration of the arrangement of dislocations in a well-annealed crystal; the Frank net. *(From Cottrell, The Properties of Materials at High Rates of Strain, Inst. Mech. Eng., London, 1957.)*

FURTHER READING

Introduction to Materials

Askeland DR, Fulay PP: *Essentials of materials science and engineering*, 2010, Nelson Publishing.

Callister WD Jr: *Materials science and engineering: an introduction*, ed 7, 2007, Wiley.

Advanced Reviews of Materials

Cahn RW, Haasen P: *Physical metallurgy*, (vols 1–3), 1996, North-Holland.

Cahn RW, Haasen P, Kramer EJ: *Materials science and technology: a comprehensive treatment*, (vols 1–18), 1993–on, Wiley-VCH.

Crystals: Atomic and Electronic Structure

Barrett CS, Massalski TB: *Structure of metals*, 1980, Pergamon.

Cullity BD, Stock SR: *Elements of X-ray diffraction*, 2003, Pearson Education.

Hammond C: *The basics of crystallography and diffraction*, ed 2, 2001, Oxford University Press.

Kelly A, Groves GW, Kidd P: *Crystallography and crystal defects*, 2000, Wiley.

Kittel C: *Introduction to solid state physics*, ed 8, 2005, Wiley.

Pettifor D: *Bonding and structure of molecules and solids*, 1995, Oxford University Press.

Sutton AP: *Electronic structure of materials*, 1993, Oxford University Press.

Tilley RJD: *Crystals and crystal structures*, 2006, Wiley.

Point Defects, Stacking Faults, Twins and Grain Boundaries

Abromeit C, Wollenberger H, editors: *Vacancies and interstitials in metals and alloys, materials science forum*, (vols 15–18), 1987, Trans Tech Publications.

Christian JW: *The theory of transformations in metals and alloys (Part I and II)*, ed 3, 2002, Pergamon.

Christian JW, Mahajan S: Deformation twinning, *Prog Mater Sci* 39:1, 1995.

Peterson NL, Siegel RW, editors: *Properties of atomic defects in metals*, 1978, North-Holland.

Smallman RE, Harris JE, editors: *Vacancies '76*, London, 1977, The Metals Soc.

Sutton AP, Balluffi RW: *Interfaces in crystalline materials*, 1995, Oxford University Press.

Dislocations

Nabarro FRN, et al, editors: *Dislocations in solids*, (vols 1–16), 1979–2010, North-Holland. (This series contains 96 chapters by experts on a wide range of dislocation properties and phenomena in materials affected by dislocations.)

Argon A: *Strengthening mechanisms in crystal plasticity*, 2007, Oxford University Press.

Cottrell AH: *Dislocations and plastic flow in crystals*, 1953, Oxford University Press.

Friedel J: *Dislocations*, 1964, Pergamon.

Hirth JP, Lothe J: *Theory of dislocations*, 1982, Wiley.

Nabarro FRN: *The theory of crystal dislocations*, 1967, Oxford University Press.

Read WT: *Dislocations in crystals*, 1953, McGraw-Hill.

Seeger A: *Plasticity of crystals, Handbuch der physik*, (vol. VII, part II, p. 1), 1958, Springer-Verlag.

Veysièrre P: Dislocations and the plasticity of crystals. In Meyers MA, Armstrong RW, Kirchner H, editors: *Mechanics and materials: fundamental linkages*, 1999, John Wiley. p. 271.

Weertman J, Weertman JR: *Elementary dislocation theory*, 1993, Oxford University Press, (1st edition 1964, Macmillan).

CHAPTER 2

Observation of Dislocations

2.1 INTRODUCTION

A wide range of techniques has been used to determine the properties of individual dislocations and study the distribution, arrangement and density of dislocations. Among the experimental techniques, transmission electron microscopy has been dominant for the past 50 years. In this method (described in the next section), dislocations are examined either singly or in groups at high magnification in specimens ~ 0.1 to $4.0\,\mu m$ thick. Four other experimental techniques (described in section 2.3) have proved useful for providing information about dislocations either in particular materials or prior to the development of transmission electron microscopy: (1) surface methods, in which the point of emergence of a dislocation at the surface of a crystal is revealed; (2) decoration methods, in which dislocations in bulk specimens transparent to light are decorated with precipitate particles to show up their position; (3) X-ray diffraction, in which local differences in the scattering of X-rays are used to show up the dislocations; (4) field ion microscopy and atom probe tomography, which reveal the position of individual atoms. Unlike (4), methods (1), (2) and (3) do not resolve the arrangement of atoms at the dislocation, but rely on such features as the strain field in the crystal around the dislocation (Chapter 4) to make it visible. Visibility in conventional transmission electron microscopy is also due to contrast resulting from strain, but the more recent technique of high-resolution transmission electron microscopy does provide for resolution at the atomic scale in thin specimens.

Computer-based simulation methods have been employed extensively in recent years and can be divided into two main groups. (1) Atomic-scale simulation, in which the atomic structure of dislocations and their interaction with other crystal defects is investigated in model crystals containing up to typically a few million atoms, i.e. size up to $\sim 100\,nm$. (2) Dislocation dynamics simulation, in which dislocations are treated as flexible lines that

Introduction to Dislocations.

move and interact within an elastic medium of size up to $\sim 100\ \mu m$. Simulation methods are described in section 2.4.

2.2 ELECTRON MICROSCOPY

General Principles

Transmission electron microscopy has been the technique most widely used for the observation of dislocations and other crystal defects. Static arrangements of defects are usually studied, but in some cases, miniaturized tensile stages are used to deform specimens (less than 3 mm in length) within a microscope, thus providing direct observation of dislocation interaction and multiplication processes. Transmission electron microscopy is applicable to a wide range of materials, subject only to the conditions that specimens can be prepared in very thin section and that they remain stable when exposed to a beam of high-energy electrons within a high vacuum. The electrons in a *conventional* transmission electron microscope have an energy of typically 100 keV. Inelastic scattering of the electrons as they pass through a solid sets an upper limit to the specimen thickness of from about 100 nm for heavy, high atomic number elements to 1000 nm for light elements. Only for *high-voltage* microscopes of energy ~ 1 MeV can these limits for electron transparency be exceeded, and then only by about five times. Like its light counterpart, the electron microscope uses lenses to focus the beam and produce an image, but the lenses are electromagnetic. The electrons behave as de Broglie waves, with a wavelength of 3.7 pm at 100 keV, and this sets a theoretical limit to the resolution of the same order of magnitude. In practice, lens aberrations and electrical and mechanical stabilities place the limit at typically 0.2 to 0.4 nm.

Schematic ray diagrams for a conventional transmission microscope are shown in Fig. 2.1. Electrons emerge from the condenser lens onto the specimen, which is usually a disc a few mm in diameter. Within a *crystalline* sample, the scattered electrons are concentrated into discrete directions. These *diffracted* beams satisfy the geometrical requirements of *Bragg's Law*, in that each set of parallel crystal

FIGURE 2.1

Two basic operations of the TEM imaging systems involving projecting (a) the diffraction pattern and (b) the image on to the viewing screen. The intermediate lens selects either the back focal plane or the image plane of the objective lens as its object.

specimen

objective lens

intermediate image 1

intermediate lens

intermediate image 2

projector lens

diffraction pattern

screen

final image

(a)   (b)

planes diffract electrons in a specific direction. The diffracted electron beams are brought to focus in the back focal plane of the objective lens (Fig. 2.1 (a)), which is the plane of the diffraction pattern. When the microscope is operated in the *diffraction mode*, the beam from the diffraction lens is focused on the back focal plane and the subsequent lenses project a magnified diffraction pattern on the fluorescent screen. The objective lens also produces an inverted image of the specimen in the first image plane, however, and if the beam from the diffraction lens is focused on this plane, the microscope is in the *imaging mode* and produces a magnified image (Fig. 2.1(b)), the final magnification being variable between about 10^2 and 10^6 times. An aperture is placed in the objective back focal plane to permit only one beam to form the image. When the beam transmitted directly through the specimen is selected, a *bright-field* image is displayed, whereas when one of the diffracted beams is chosen, a *dark-field* image is observed. For the analysis of defects observed in the imaging mode, information on the diffraction conditions is required and is obtained in the diffraction mode. Of particular importance is the *diffraction vector* **g** perpendicular to the diffracting planes.

The image simply reveals the variation in intensity of the selected electron beam as it leaves the specimen. In the theory of this diffraction contrast, it is usually assumed for simplicity that the intensity variation across the bottom surface of the specimen can be obtained by considering the specimen to consist of narrow columns (a few nm wide) with axes parallel to the beam: the electron intensity does not vary across a narrow column and is independent of neighboring columns. The final image is therefore an intensity map of the grid of columns. Furthermore, a two-beam condition is assumed to apply, wherein the contribution of the electrons in the directly transmitted beam and *one* diffracted beam only, is considered, i.e. only one set of crystal planes is at, or close to, the Bragg angle. These *column* and *two-beam approximations* have been shown to be valid for conventional microscopy. If the specimen is perfectly flat, uniformly thick and free of defects, the image is homogeneous with no variations in intensity. Image contrast only arises if variations in beam intensity occur from one part of the specimen to another. For example, if the specimen is bent in one region sufficiently to affect crystal planes which are close to the Bragg condition, strong variations in intensity can result. Dislocations and many other crystal defects bend crystal planes.

The effect of crystal defects on the image depends on the vector **u** by which atoms are displaced from their perfect lattice sites. In the column approximation, **u** within a column varies only with z, the coordinate along the column axis. Solutions of the equations for the electron intensities contain a factor $\mathbf{g} \cdot \mathbf{u}$, or alternatively $\mathbf{g} \cdot d\mathbf{u}/dz$, which is not present for a perfect crystal. Thus, diffraction conditions for which $\mathbf{g} \cdot \mathbf{u}$ (or $\mathbf{g} \cdot d\mathbf{u}/dz$) is zero will not lead to contrast in the image. The physical interpretation of these effects follows from the fact that du/dz is the variation of displacement with depth and it therefore measures the bending of atomic planes. If the reflecting planes normal to **g** are not tilted by the defect, there is no change in diffracted intensity and the defect is invisible.

Dislocations

It can be seen from Fig. 1.18 that certain planes near a dislocation are bent, and the bending decreases with increasing distance from the dislocation. (Expressions for the elastic field of a straight dislocation are given in section 4.3.) If these planes are bent into a strongly diffracting orientation, the intensity of the directly transmitted beam will be reduced (and that of the diffracted beam increased) in columns near the dislocation. This is shown schematically in Fig. 2.2. The dislocation will appear as a dark line in the bright-field image (or as a bright line in a dark-field image). An example of two rows of dislocations is shown by the bright-field micrograph of Fig. 2.3. The actual dislocation structure observed in a particular specimen depends very much on the material and, if it has been deformed, on the strain, strain rate and temperature it has experienced. Examples are to be found in Figs 2.4, 6.17, 7.21, 9.1, 9.2, 9.9, 10.12 and 10.18. The width of a dislocation image is determined by the variation of du/dz across the specimen, i.e. by the range of the dislocation strain field, and is typically 10 to 50 nm. The actual form of the image depends on the diffraction conditions, the nature of the dislocation and its depth in the foil. It may appear as a single line (not necessarily centered on the real dislocation), a double line, a wavy line or a broken line. Also, the line may be invisible, as explained above, and this may be exploited to determine the Burgers vector **b**.

The *invisibility criterion* is $\mathbf{g} \cdot \mathbf{u} = 0$. Consider first a straight screw dislocation. In an isotropic medium, planes parallel to the line remain flat (see Fig. 1.18), for **u** is parallel to **b** (section 4.3). Hence, when **g** is perpendicular to **b**, $\mathbf{g} \cdot \mathbf{u} = \mathbf{g} \cdot \mathbf{b} = 0$ and the invisibility criterion is satisfied. If the specimen is tilted in the microscope to find two sets of diffracting planes $\mathbf{g}_1$ and $\mathbf{g}_2$ for which the line is invisible, **b** must be perpendicular to both $\mathbf{g}_1$ and $\mathbf{g}_2$ and therefore has the direction of $\mathbf{g}_1 \times \mathbf{g}_2$. For an edge dislocation, all planes parallel to the line are bent, and **u** is non-zero in all directions perpendicular to the line. The criterion $\mathbf{g} \cdot \mathbf{u} = 0$, therefore, requires in this case that both $\mathbf{g} \cdot \mathbf{b}$ and $\mathbf{g} \cdot (\mathbf{b} \times \mathbf{t})$ be zero, where **t** is a vector along the line. It is satisfied only when **g** is parallel to the line, for only planes perpendicular to the line remain flat (see Fig. 1.18). For the mixed dislocation, which produces a combination of edge and screw displacements, there is no condition for which $\mathbf{g} \cdot \mathbf{u}$ is exactly zero. Also, in anisotropic crystals no planes

FIGURE 2.2

Planes near an edge dislocation bent into the orientation for diffraction.

Incident

Transmitted   Diffracted

remain flat around edge and screw dislocations, except in a few special cases. Thus,
the invisibility method for determining **b**
often relies on finding diffraction vectors
which result in weak contrast rather than
complete invisibility. It has proved, nevertheless, a powerful technique in dislocation
studies. An example is illustrated in Fig. 2.4.
Here, three different diffraction vectors **g**
have been chosen to produce three different
images of the same field of view. It contains
a network of four sets of dislocation lines.

Although modern transmission electron
microscopes have resolution limits of less
than 1 nm, the dislocation image width is
much greater than this, and fine structure of
the dislocation, such as splitting into partials (Chapters 5 and 6) may not be
resolved. The *weak-beam* technique can overcome this disadvantage. It utilizes the fact
that when the specimen is tilted away from
the Bragg orientation the diffracted beam
from the perfect crystal is weak; strong dif-
fraction from a dislocation can only occur when $\mathbf{g} \cdot (d\mathbf{u}/dz)$ is large. The
strong bending of planes this requires only occurs close to the center of the
dislocation. The weak-beam images of dislocations are therefore much nar-
rower (typically 1–2 nm) than those obtained under strong-beam conditions, although the experimental procedure is more demanding. An example
of closely-spaced dislocations resolved in silicon is shown in Fig. 2.5. The
invisibility criteria are the same as those discussed above, and the weak-
beam method has been applied to many materials.

FIGURE 2.3

(a) Thin film transmission
electron micrograph
showing two parallel rows
of dislocations. Each dark
line is produced by a
dislocation. The
dislocations extend from
top to bottom of the foil
which is about 200 nm
thick. (b) The line diagram
illustrates the distribution
of the dislocations in the
foil and demonstrates that
the photograph above
represents a projected
image of a three-
dimensional array of
dislocations.

Planar Defects

Stacking faults and other coherent interfaces without long-range strain fields
also form characteristic image contrast in the transmission electron microscope. A fault is produced by the displacement of the crystal above the fault
plane by a *constant* vector **u** relative to the crystal below (section 1.3). For
columns passing through the fault, this therefore introduces a phase factor
$2\pi \, \mathbf{g} \cdot \mathbf{u}$ into the main and diffracted beam amplitudes. Thus, when $\mathbf{g} \cdot \mathbf{u}$ is
zero or an integer, no contrast is present and the fault is invisible. This is
demonstrated schematically in Fig. 2.6(a), where **u** lies in the plane of the
fault and the diffracting planes are parallel to **u**. For other values of $\mathbf{g} \cdot \mathbf{u}$,

FIGURE 2.4

Illustration of the use of the $\mathbf{g} \cdot \mathbf{b} = 0$ method to determine the Burgers vector of dislocations in an Al–2% Mg alloy. *(From Lindroos, Phil. Mag. 24, 709, 1971.)*

however, the fault produces contrast, as implied by Fig. 2.6(b). When the fault is inclined to the specimen surface, contrast takes the form of light and dark fringes parallel to the line of intersection of the fault plane with the surface, as shown by the example in Fig. 2.7. The precise form of the fringe contrast depends on the diffraction conditions employed, and this enables faults and defects such as coherent twin, grain and phase boundaries to be analyzed.

Lattice Imaging

Under certain conditions, it is possible to study the atomic structure of defects by exploiting the full resolution of the transmission electron microscope to image the actual columns of atoms in a thin sample. An edge dislocation lying normal to the specimen surface, for example, produces an extra half-plane in the image, as shown by the example in Fig. 9.21. Other experimental images are shown in Figs 9.27(b) and 9.29. Unlike conventional microscopy in which the image is formed by a single (direct or diffracted) beam, lattice imaging requires two or more beams to pass through the objective aperture in order that the periodicity of the object, i.e. the lattice planes, be resolved. Spacings down to 0.1 nm have been resolved by this technique. In early studies, it was assumed that one-to-one correspondence exists between the fringes and the crystal planes, and that atomic positions can be visualized directly from the image. It is now known, however, that this is only true in special circumstances. They require the specimen to be very thin (~1–10 nm) so that rediffraction of diffracted beams cannot occur. Interpretation of lattice images is difficult if this condition is not met.

Image Simulation

The transmission electron microscope image of a dislocation depends not only on the diffraction conditions but also on the form of the dislocation and the nature of the specimen. For example, the $\mathbf{g} \cdot \mathbf{b} = 0$ invisibility criterion only applies generally to isotropic crystals, which have the same elastic properties in all directions. Except in a few cases, real crystals are elastically anisotropic to varying degrees, and when the anisotropy is strong the criterion cannot be applied to analyze dislocations. Another example arises when the dislocation is a loop, as occurs in irradiation damage or quenching. If the loop is too small (less than about 10 nm in diameter) to be resolved distinctly, conventional analysis is not possible, although the image does contain information on loop plane and Burgers vector. In these situations, defect identification is best undertaken by matching real micrographs to computer-generated ones. The latter are obtained for all possible forms the unknown defect might have using the same diffraction conditions as those employed experimentally. It is possible to identify the defect uniquely by suitable choice of conditions.

FIGURE 2.5

A weak-beam 220 dark-field image of a dislocation in an annealed silicon specimen showing constricted segments. The inset shows the diffracting conditions used to form the image. *(From Ray and Cockayne, Proc. Roy. Soc. A325, 543, 1971.)*

An example of real and simulated images due to small vacancy loops in ruthenium is shown in Fig. 2.8. Each simulated micrograph is produced by first computing the electron intensity using the column approximation, with the displacement field of the trial defect obtained from elasticity theory. Typically, several thousand columns are required. The resulting grid of intensities is then used to generate a corresponding grid of dots, the size (or 'greyness') of each dot being given by the intensity. The simulated image is therefore formed in a manner analogous to photograph reproduction in newspapers.

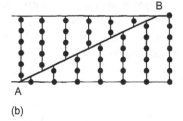

FIGURE 2.6

Displacements of the reflecting planes shown by lines at a stacking fault *AB*. Diffraction vector **g** is perpendicular to these planes. In (a) $\mathbf{g} \cdot \mathbf{u} = 0$ and no contrast occurs. In (b) $\mathbf{g} \cdot \mathbf{u} \neq 0$ and interference between waves from above and below the fault gives a fringe pattern. *(From Howie, Metallurgical Reviews, 6, 467, 1961.)*

Another example is provided in Fig. 2.9. In Fig. 2.9(a), atomic positions around a stacking-fault tetrahedron (section 5.7) formed from 45 vacancies in copper have been calculated by atomic-scale computer simulation (section 2.4). The atomic coordinates have been used to predict the weak-beam image in a

(a)   0.1 μm

Stacking fault

(b)

FIGURE 2.7

(a) Thin film transmission electron micrograph showing the fringe pattern typical of stacking faults which lie at an angle to the plane of the foil, as illustrated schematically in (b).

transmission electron microscope (Fig. 2.9(b)). It matches the experimental image (Fig. 2.9(c)) from proton-irradiated copper very well.

Other Factors

Some factors which limit the applicability of transmission electron microscopy have been noted already. To attain the specimen thickness required for electron transparency, it is usually necessary to first cut the samples by mechanical or spark-erosion machining to a thickness of about 10–100 μm, and then thin the region in the middle by techniques such as electrolytic polishing, chemical polishing or ion-beam sputtering. Other methods, such as mechanical cleavage for layer crystals or deposition from solution for some organic crystals, have been used less commonly. Some materials, such as organic solids, are unstable under the effect of the electron beam and can only be viewed for short times before decomposing.

For transmission electron microscopy to be of value, it is important that the dislocations seen in the thin foil sample are typical of those of the bulk material. This is sometimes not the case. For example, inadvertent deformation during thinning can introduce additional dislocations. On the other hand, dislocations close to the surface experience forces attracting them to the surface (section 4.8), and this effect can substantially reduce the dislocation density in thin specimens. As explained earlier, high-voltage microscopes allow thicker samples to be used so that these problems are reduced. However, this advantage has to be offset against the greater cost of the instrument. Also, if an atom in a crystal acquires a kinetic energy greater than its threshold *displacement energy*, which is typically 25 eV, it can escape from its lattice site and create a vacancy-interstitial pair. Materials with atomic weight up to about 170 suffer such displacement damage from bombardment with electrons with energy of 1 MeV or greater, and their structure is therefore changed by examination in the high-voltage microscope. This feature can, of course, be used with advantage for radiation-damage studies.

2.3 OTHER EXPERIMENTAL METHODS

Surface Methods

If a crystal containing dislocations is subjected to an environment which removes atoms from the surface, the rate of removal of atoms around the point at which a dislocation emerges at the surface may be different from that for the surrounding matrix. The difference in the rate of removal arises from one or more properties of the dislocation: (1) lattice distortion due to the dislocation; (2) geometry of planes threaded by a screw dislocation, which produces a surface pit; (3) concentration of impurity atoms at the dislocation, which changes the local chemical composition. If the rate of removal is more rapid around the dislocation, a pit is formed, Fig. 2.10, and if less rapid a small hillock is formed. The most common methods for the slow, controlled removal of atoms

FIGURE 2.8
Real and computer-simulated images of vacancy loops produced by heavy-ion bombardment of ruthenium. The diffraction vector is **g**, and from image matching, loops α and β have the Burgers vector **b** and loop normals **n** shown. *(Courtesy of W. Phythian.)*

are chemical and electrolytic etching. Other methods include thermal etching, i.e. evaporation, and sputtering, in which surface atoms are removed by gas ion bombardment. The specimen surface in Fig. 2.11 has been etched and preferential attack has produced black spots identified as pits.

Note that when a screw dislocation emerges at the surface of a crystal that is growing due to deposition of atoms from a liquid or vapor, the surface step (Fig. 2.10(c)) provides a favorable location for atom attachment. The screw dislocation mediates growth.

Confirmation of a *one-to-one correspondence* between dislocations and etch pits is illustrated in Fig. 2.12. The photograph shows a regularly spaced row of etch pits formed at the boundary between two germanium crystals. Precise X-ray measurements to determine the misorientation between the crystals showed that the boundary was a symmetrical pure tilt boundary (section 9.2), which can be described as an array of edge dislocations spaced one above the other as shown in Fig. 2.12(b). Such an array produces a tilt between the grains on opposite sides of the boundary. If b is the magnitude of the Burgers vector of the edge dislocations and D their distance apart, the angle of misorientation is $\theta = b/D$ (equation (9.2)). For the boundary in Fig. 2.12(b), the measured value of θ was 65 sec of arc and the predicted value of D was 1.3 μm. This agrees closely with the spacing of the etch pits shown in the photograph, confirming that in this example there is a one-to-one correspondence between etch pits and dislocations.

Although etch pit studies are limited to the surface examination of bulk specimens the technique can be used to study the movement of dislocations as

FIGURE 2.9

Stacking-fault tetrahedron in irradiated copper. (a) Atoms in two {111} planes through a tetrahedron in a computer simulation. (b) Experimental and (c) simulated weak-beam images, with a wire frame on the right showing the orientation of the defect. *(From Schaublin, Dai, Osetsky and Victoria (1998), Inter. Congr. on Electron Microscopy, vol. 1, p. 173, Institute of Physics Publishing.)*

demonstrated by Gilman and Johnston (1957) in lithium fluoride. This is illustrated in Fig. 2.13. The site of a stationary dislocation in a crystal appears as a sharp bottom pit. If a dislocation moves when a stress is applied to the crystal, the distance moved depends on the applied stress and the length of time the stress is applied. If the crystal is etched again, the new position of the dislocation is revealed by a new sharp bottom pit. The etchant also attacks the old pit which develops into a flat bottom pit.

In general, surface techniques are limited to crystals with a low dislocation density, less than $10^{10} \, \text{m}^{-2}$, because the etch pits have a finite size and are very difficult to resolve when they overlap each other. The distribution of dislocations in three dimensions can be obtained, using this technique, by taking successive sections through the crystal.

Decoration Methods

The dislocations in crystals which are transparent to visible light and infrared radiation are not normally visible. However, it is possible to *decorate* the dislocations by inducing precipitation along them. Sites along the dislocation are favored by the lattice distortion there (section 10.5). The effect produced is similar in appearance to a row of beads along a fine thread. The position of the dislocation is revealed by the scattering of the light at the precipitates and can be observed in an optical microscope. In most applications of this method the decoration process involves the heating of the crystals before examination and this restricts the use of the method to the study of 'recovered' or high-temperature deformation structures. It is not suitable for studying structures formed by low-temperature deformation.

The procedure with the widest range of application, which has been used to study dislocations in alkali halide and semiconductor crystals, is to dope with impurity atoms. Precipitation can subsequently be induced along the dislocations by suitable heat treatment. Figure 2.14 illustrates this for KCl to which a small amount of AgCl was added. The crystals were annealed to precipitate silver particles on the dislocations. The helical dislocation in CaF$_2$ shown in Fig. 3.21 was revealed by the same technique. Dislocations in silicon can be decorated by a similar technique and then observed with infrared radiation by diffusing in metallic elements such as copper or aluminum. These decoration techniques can be used in combination with etch-pit studies to demonstrate the one-to-one correspondence between dislocations and etchpits. They have been reviewed by Amelinckx (1964).

An optical method which does not depend on decoration has been used to study dislocations in magnetic bubble-domain materials. It utilizes detection by polarised light of the *birefringence* induced by the stress from the dislocation. A brief review is given by Humphreys (1980).

X-ray Diffraction Topography

Direct observation of dislocations with X-rays is achieved by a method somewhat similar to electron diffraction, but with a greatly reduced resolution. This rather specialized technique produces dislocation image widths of 1 μm or greater. Consequently, it is applicable only to

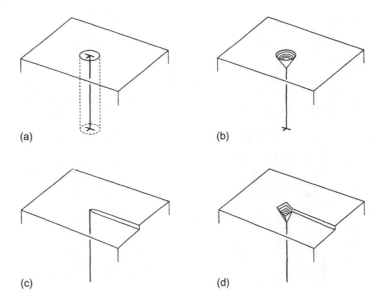

(a)   (b)

(c)   (d)

FIGURE 2.10

Formation of etch pits at the site where a dislocation meets the surface. (a) Edge dislocation, the cylindrical zone around the dislocation represents the region of the crystal with different physical and chemical properties from the surrounding crystal. (b) Conical-shaped pit formed at an edge dislocation due to preferential removal of atoms from the imperfect region. (c) Emergent site of a screw dislocation. (d) Spiral pit formed at a screw dislocation; the pits form by the reverse process to the crystal growth mechanism.

20μm

FIGURE 2.11

Etch pits produced on the surface of a single crystal of tungsten. *(From Schadler and Low, unpublished.)*

(a)

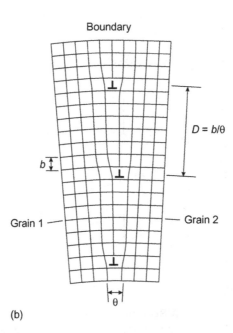

(b)

FIGURE 2.12

(a) A row of etch pits formed at the boundary between two germanium crystals. The etch pits are uniformly spaced. (b) Diagrammatic representation of the arrangement of dislocations in the boundary revealed by the etch pits in (a). This is a symmetrical pure tilt boundary which consists of a vertical array of edge dislocations with parallel Burgers vectors of the same sign. *(After Vogel, Pfann, Corey and Thomas, Physical Review 90, 489, 1953.)*

the study of crystals with low dislocation densities $\leqslant 10^{10}\,\mathrm{m}^{-2}$, but has the advantage that the penetration of X-rays is greater than electrons so that much thicker specimens can be used. The specimen, usually a large single crystal, is oriented with respect to the X-ray beam so that a set of lattice planes is set at the Bragg angle for strong reflection. The reflected beam is examined photographically. As in electron diffraction, any local bending of the lattice associated with a dislocation results in a change in the reflection conditions and the X-rays are scattered differently in this region. The difference in the intensity of the diffracted X-rays can be recorded photographically. A photograph of dislocations in a silicon crystal revealed by X-ray diffraction topography is shown in Fig. 2.15.

Field Ion Microscopy and Atom Probe Tomography

The maximum resolution of the electron microscope does not, in general, allow the examination of the positions of individual atoms and, in particular, point defects cannot be detected unless they form in clusters. This limitation is overcome by the field ion microscope which has a resolution of 0.2 to 0.3 nm. The specimen is a fine wire which is electropolished at one end to a sharp hemispherical tip. Within the microscope chamber, the tip is sharpened further to a radius between 5 and 100 nm by *field evaporation*, a process in which a high electric field ($\sim 100\,\mathrm{V\,nm}^{-1}$) removes atoms from the tip surface.

An image of the atom positions on the tip surface is obtained by *field ionization*. The specimen is contained at cryogenic temperatures in a chamber evacuated to a high vacuum and into which a low-pressure imaging gas of helium or neon is admitted. The specimen is held at a positive potential of several kV with respect to a fluorescent screen, thereby producing a high field in the range $10-100\,\mathrm{V\,nm}^{-1}$ at the specimen tip. The field polarizes gas atoms near the tip and attracts them towards it. After repeated collisions with the cold surface, the gas atoms slow down sufficiently for them to occasionally become ionized as they pass through the strong field at the protruding surface atoms. The electrons enter the specimen and the positive gas ions are repelled by the field to the screen. Since they travel approximately radially, there is a geometrical magnification of about d/r, where d is the distance from the tip to the image and r is the tip radius. Magnifications of the order of 10^6 times are obtained. Only about 1 in 10 of the surface atoms protrude sufficiently to ionize the image gas, and they appear as bright spots in the image. These atoms are at the edge of the crystal planes where they intersect the tip surface. The image therefore consists of sets of concentric rings of bright spots, each set corresponding to the surface ledges of planes of a particular crystallographic index. A vacancy is revealed in the image as a missing spot, and more complicated defects produce greater disruption to the ring pattern. An example of a grain boundary in tungsten is shown in Fig. 2.16. It is seen that the disruption is strongly localized along the boundary, rather like the simple bubble-raft model of Fig. 1.14.

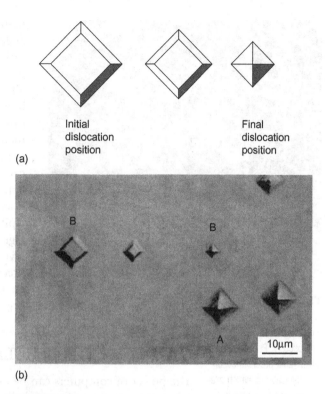

(a)

(b)

FIGURE 2.13

Etch pits produced on a lithium fluoride crystal. The crystal has been etched three times. The dislocation at A has not moved between each etching treatment and a large pyramid-shaped pit has formed. The dislocation revealed by the three pits B moved from left to right between etching treatments to the positions indicated by the pits. Subsequent etching of a pit after the dislocation has moved produces a flat bottom pit. *(From Gilman and Johnston (1957),* Dislocations and Mechanical Properties of Crystals, *p. 116, Wiley.)*

Although the image only shows details of the positions of atoms on a surface, field ion microscopy may be used to reveal three-dimensional structures. This is achieved by alternating the formation of field ionization images with the repeated removal of surface atoms by field evaporation. It has been used to investigate structures associated with radiation damage, dislocations and grain boundaries.

As an enhancement of the field ion technique, atom probe field ion microscopy has been developed to investigate the distribution of individual alloying and impurity atoms in materials. In early studies, an aperture at the image screen was used to permit a surface atom removed by field evaporation to enter a mass spectrometer, where its nature was analyzed. The

FIGURE 2.14
A thin crystal of KCl examined in an optical microscope. Particles of silver have precipitated on the dislocations, which are in the form of a network. Only part of the network is in focus. *(From Amelinckx, Acta Metall. 6, 34, 1958.)*

material volume analyzed was restricted by an aperture size ~1 nm. By the end of the 1990s, three-dimensional atom probes (3DAPs) using position-sensitive detectors for single atoms allowed atoms from regions ~10 nm in size to be collected, and more recent developments in this field of *atom probe tomography* have increased the size again to over 50 nm, so that volumes containing over 100 million atoms can be analyzed. The technique has been applied to a wide range of materials to study microstructural/chemical features such as pre-cipitation and segregation at the atomic-level. Figure 2.17 shows orthogonal 3DAP images revealing segregation of phosphorus atoms and precipitation of copper along a dislocation in an iron-based alloy. The formation of *Cottrell atmospheres* of solute atoms along dislocations is discussed in section 10.5.

2.4 COMPUTER SIMULATION

The power of computers can be exploited to investigate properties of disloca-tions that are not open to direct study by experiment. The techniques involved can be divided into two broad categories. One simulates disloca-tions and other defects in crystals modeled at the atomic scale. The other treats the crystal as an elastic continuum with no atomic discreteness.

Atomic-Level Simulation

Generally, the experimental techniques described above do not permit the atom positions at the center or *core* of a dislocation to be studied. (The *core* is the region around a dislocation in which the distortion is too large to be described by elasticity theory (see section 4.4): it is typically a few inter-atomic spacings in size.) The core structure is important because, as will be seen in later chapters, it determines characteristics such as the slip plane of the dislocation, the ease with which it slips, its interaction with other defects and its effect on electrical behavior of non-metals. These features can be investigated by computer simulation.

Two distinct approaches are used. *First principles* or *ab initio* methods incor-porate the quantum mechanical nature of bonding between atoms. The energy and configuration of the atomic system is determined by solving Schrödinger's equation for interacting electrons. Several computer codes based on density functional theory are available for this purpose, but they are demanding in computer power and, for that reason, are restricted to model systems containing typically 100 to 1000 atoms. They have been used to determine the local atomic structure and energy associated with solute atoms, vacancies, self-interstitials and small clusters (up to four or five) of these defects. They are also valuable for providing a database of crystal properties for fitting the adjustable parameters in empirical interatomic potentials (see below). First principles calculation does not lend itself easily to simulation of dislocations, however, because the small model size is not generally compatible with the long-range strain field of these defects (Chapter 4). Nevertheless, important information on some dislocations has been obtained by choosing specific model shape and boundary conditions perpendicular to the dislocation in order to

20mm

FIGURE 2.15

X-ray diffraction topograph (micrograph) showing dislocations in a single crystal of silicon. No magnification occurs in recording the topograph, but by using very fine grain sized photographic emulsions subsequent magnification up to about × 500 is possible. *(From Jenkinson and Lang (1962), Direct Observation of Imperfections in Crystals, p. 471, Interscience.)*

minimize this effect. Also, use of periodic boundary conditions parallel to the dislocation allows an infinitely long dislocation to be simulated. Notably, this method has been used to establish that the atomic structure of the core of the screw dislocation in iron and some other body-centered cubic metals has the non-degenerate form described in section 6.3 (Frederiksen and Jacobsen, 2003).

The other approach to atomic-scale simulation employs models that provide a reasonably accurate description of the atomic structure of the dislocation core yet are large enough to allow for the influence of the dislocation strain field. Use of these larger models enables effects of applied stress and temperature on dislocation properties, and the interaction of dislocations with other defects to be investigated. The methods are broadly known as *molecular statics* and *molecular dynamics*. In both cases, a computer program is used first to generate atomic coordinates for a perfect crystal of specified structure, orientation and size, and then a dislocation is introduced by appropriate displacement of the atoms. An infinitely long dislocation is simulated by use of periodic boundary conditions in the direction of the dislocation line. Other defects such as point defects, precipitates, dislocation loops and grain boundaries can also be created if required. Motion of the dislocation can be studied by applying shear displacements or forces to atoms on the boundaries of the model. Visualization is an important aspect of computer simulation because it enables dislocation behavior to be more clearly understood. Several algorithms are in use to identify the position and coordination of atoms in the dislocation core.

FIGURE 2.16

Field ion micrograph of a grain boundary at the tip of tungsten needle. Each bright spot represents a tungsten atom. *(From Scientific American, Sept. 1967.)*

The potential energy of the model system and the forces between the atoms are computed from an interatomic potential. Potentials are available for the simulation of many materials. Those used for metals are based mainly on the Embedded Atom Model (EAM). Their empirical parameters are obtained by fitting to material properties such as elastic constants, phonon spectra, cohesive energy, stacking fault energy, and surface and point defect energies, some of which are derived from *ab initio* calculations.

In molecular statics, a crystal at temperature $T = 0$ K is modeled, i.e. the kinetic energy of the atoms is equal to zero, and the system achieves equilibrium when the potential energy is minimized. Molecular statics codes find this state by moving the atoms repeatedly in a series of iterations. In molecular dynamics, kinetic energy is not zero and at equilibrium the average kinetic energy per atom equals $3kT/2$, where k is Boltzmann's constant. At a given instant of time t, the acceleration of every atom is calculated from the force on it due to its neighbors using Newton's second law (force = mass times acceleration). The computer code uses this information to predict the position of all atoms at time $(t + \Delta t)$, where Δt is the molecular dynamics time-step. This procedure is repeated to enable the trajectory of atoms to be followed for as long as is necessary to complete the process under investigation. To maintain accuracy, Δt is typically of the order of one to a few femtoseconds (1 fs = 10^{-15} s).

The need to achieve results in reasonable computing time limits molecular statics and molecular dynamics modeling in several ways. First, the computing time per atom is determined by the time needed to compute the energy of an atom and the force on it due to its neighbors. Thus, to minimize the time, the range of the interatomic potential should be as short as possible: the EAM potentials satisfy this requirement. Second, the total time is proportional to the number of atoms and the number of iterations (molecular statics) or time-steps (molecular dynamics) required to complete the simulation. Unless computer systems with many processors in parallel are available, models are limited to 10^5 to 10^7 atoms, but this is usually sufficient for the study of single dislocations and their interaction with other crystal defects mentioned above. With regard to time-scale in molecular dynamics, the number of time-steps that can be accomplished within a reasonable computing time on a single processor is typically in the range 10^5 to 10^7. As a consequence, the total simulated time is limited to the order of nanoseconds (1 ns = 10^{-9} s) and the applied strain rate is high compared

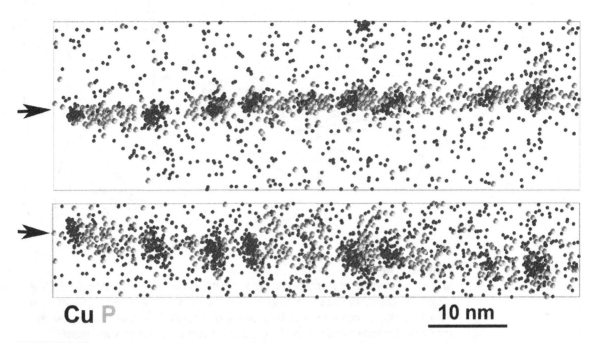

Cu P   10 nm

FIGURE 2.17

Orthogonal views of an atom map of phosphorus segregation and copper-rich precipitates on a dislocation (arrowed) in a neutron-irradiated ferritic steel. *(From Miller and Russell, J. Nucl. Mater.* **371**, *145, 2007. Copyright (2007) with permission from Elsevier.)*

with laboratory tensile tests. For example, a strain rate of 10^7 s^{-1} is required to achieve a strain of 1% in 1 ns. This corresponds to a dislocation velocity of 10 to 100 ms^{-1} in models of typical size. (The relationship between dislocation velocity, dislocation density and strain rate is derived in section 3.9.)

An example of a model is illustrated in Fig. 2.18(a). The dimensions are usually in the range 10 to 100 nm. Periodic boundary conditions are used along the direction of the dislocation (x-axis for the screw and y-axis for the edge). A shear stress τ can be applied by imposing either shear forces or displacements to atoms on the $\pm z$ boundaries, and this induces the dislocation to move on the x-y glide plane where it encounters an obstacle of interest. The shape (visualized by atoms in the core) of a dislocation overcoming a periodic row of 2 nm voids of spacing 41 nm in a model of iron at two temperatures is shown in Fig. 2.18(b). The edge dislocation with Burgers vector $\frac{1}{2}$[111] was initially straight and glided to the right under increasing applied strain. It was pinned by the voids and the shapes in the figure are those just before it broke away at the critical stress τ_c indicated. The shapes for 0 K and 300 K were obtained by molecular statics and molecular dynamics simulation, respectively. The critical stress is seen to decrease, and the dislocation to bow out less, with increasing temperature.

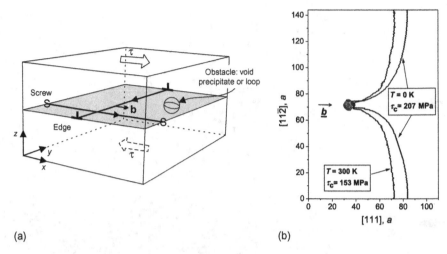

(a)   (b)

FIGURE 2.18

(a) Schematic illustration of the crystal cell used to simulate interaction of a dislocation with a row of obstacles. (b) Critical line shape for a dislocation passing a row of 2 nm voids with spacing 41 nm in iron at 0 and 300 K. Scale unit a is the lattice parameter of iron (0.286 nm). *(From Bacon, Osetsky and Rodney, Dislocations in Solids, vol. 15, p. 3 (ed. J. P. Hirth and L. Kubin), North-Holland. Copyright (2009) with permission from Elsevier.)*

The results in Figs 2.9(a), 3.13, 5.4, 5.8, 6.9, 6.10, 6.11 and 6.21 were produced by atomic-scale computer simulation. Ways in which information on the atomic core structure of dislocations obtained by simulation can be presented graphically are discussed in section 10.3.

Continuum-Level Simulation

The discrete atomic structure of a crystal is neglected in elasticity theory and the material is treated as a continuum. As will be seen in Chapter 4, the distortion created in crystals by dislocations is predicted with reasonable accuracy by linear elasticity theory and the forces on dislocations that make them move in response to external and internal stress can also be calculated in the same framework. Later chapters will demonstrate that elastic theory of straight dislocations forms the bedrock of much of our understanding of dislocation-related properties of materials. Nevertheless, dislocations are seldom straight, as has been seen already in this chapter, and there are situations where line curvature is important. Unfortunately, it is generally not possible to obtain analytical solutions for properties of curved dislocations in elasticity theory and a numerical treatment has to be used. This is the approach of the simulation method known as *dislocation dynamics* (also known as discrete dislocation dynamics and line dislocation dynamics).

Figure 2.19 illustrates the way in which dislocations are represented by nodes (drawn as circles) connected to each other by straight dislocation segments. Each segment is characterized by its Burgers vector and direction in space between two nodes, and an explicit sign convention has to be used so that Burgers vector is conserved at every node, as described in section 1.4. The degree to which segmented dislocations mimic real dislocations depends principally on two factors, as follows.

FIGURE 2.19
Representation of flexible dislocations by nodes (shown as circles) connected by straight segments.

One is the way in which nodes move in response to forces on them from their own line, other dislocations, other sources of internal stress and externally applied stress. The glide force is related to the stress by equation (4.27) and the stress at a node due to adjacent and more distant segments is determined in dislocation dynamics within the approximation of linear elasticity theory. Force arising from core energy (section 4.4) can also be included. The computation involved is non-trivial and several methods have been developed to provide computational efficiency (see Bulatov and Cai (2006) in Further Reading). The dislocation will want to change its shape if the force on a node is unbalanced. One way of allowing for this is to give the node a velocity which is proportional to the force and inversely proportional to a drag coefficient, as in equation (3.5). Realism can be introduced by setting the drag coefficient for climb (in response to a force component out of the glide plane) to be much larger than that for glide, and allowing the coefficient for glide to vary with dislocation line direction between the edge and screw orientations. The effects of temperature can also be incorporated with a temperature-dependent drag coefficient and by probability-determined waiting times for thermally-activated processes (section 10.2).

The other important issue is concerned with the density of nodes that define the dislocation line. Computational efficiency requires the number of nodes in the system to be as small as possible, but accurate representation of the line shape is achieved if the nodal density is large. It is clear from Fig. 2.19 that the density needs to be high in regions where the line curvature is strong and that it can be low when the curvature is small. Thus, the computer code has to be able to add and remove nodes in response to shape changes of dislocations during a simulation. Furthermore, segments on the same or different dislocations can react to either annihilate or form segments with a different Burgers vector, and again nodes have to be added or removed.

In some respects, dislocation dynamics is similar to molecular dynamics in that objects (atoms or nodes) move in a chosen time-step according to the

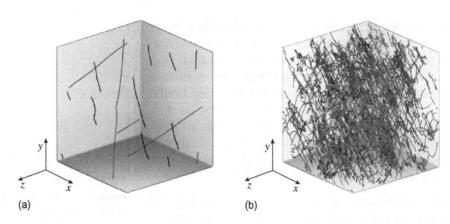

(a) (b)

FIGURE 2.20
(a) Initial dislocation arrangement for the DD simulation of a molybdenum single crystal under applied uniaxial strain. (b) Dislocation microstructure when the strain reaches 0.3%. *(From Bulatov and Cai (2006),* Computer Simulations of Dislocations, *Oxford University Press. With permission from Oxford University Press (www.oup.co.uk).*

force on them. However, dislocation dynamics has more complexity. First, the response of a node depends on the direction of the force on it. Second, the number of atoms is constant in molecular dynamics, whereas the changing number of nodes in dislocation dynamics requires that the inventory of nodes, nodal coordinates and Burgers vectors be updated after every time-step. The choice of boundary conditions can also be problematic. Articles listed under Further Reading provide more detail.

Dislocation dynamics has been applied to situations where either a few or many dislocations are involved. Interactions between two dislocations that are important in work hardening (section 10.8) provide an example of the former. As an example of the latter, Fig. 2.20 shows the dislocation arrangement in a simulation of a tensile test on a single crystal of molybdenum, a body-centered cubic metal. The model has axes in <100> directions and size $10 \times 10 \times 10 \ \mu m^3$. The initial dislocation arrangement shown in Fig. 2.20(a) was chosen to consist of lines orientated preferentially along the <111> screw orientations. Figure 2.20(b) shows the dislocations in the computational cell after it had been strained by 3% in the x direction at a rate of $1 \ s^{-1}$. The large increase in dislocation density is apparent. The computing demands of this simulation were such that it used a parallelized code on computer clusters in which the number of processors increased from 16 to 512 as the dislocations, and therefore number of nodes, multiplied. The plots of stress and dislocation density versus strain for this simulation are described in section 10.8 (Fig. 10.26).

FURTHER READING

Experimental Techniques

Amelinckx S: The direct observation of dislocations, *Solid State Phys*, Supplement 6, 1964, Academic Press.

Amelinckx S, Gever R, van Landuyt J, editors: *Diffraction and imaging techniques in materials science*, ed 2, 1978, North-Holland.

Anstis GR, Hutchinson JL: High-resolution imaging of dislocations. In Nabarro FRN, editor: *Dislocations in Solids*, 1992, North-Holland. vol 9, chap. 44.

Brandon D, Kaplan WD: *Microstructural characterization of materials*, 2 ed, 2008, Wiley.

Coupeau C, Girard J-C, Rabier J: Scanning probe microscopy and dislocations. In Nabarro FRN, Hirth JP, editors: *Dislocations in solids*, vol 12, 2004, North-Holland, p. 273.

Egerton R: *Physical principles of electron microscopy: an introduction to TEM, SEM, and AEM*, 2005, Springer.

Goodhew PJ, Humphreys J, Beanland R: *Electron microscopy and analysis*, 3 ed, 2001, Taylor and Francis.

Humphreys CJ: Imaging of dislocations. In Nabarro FRN, editor: *Dislocations in solids*, vol 5, 1980, North-Holland, p. 1.

Ice GE, Barabash RI: White beam microdiffraction and dislocations gradients. In Nabarro FRN, Hirth JP, editors: *Dislocations in solids*, vol 13, 2007, North-Holland, p. 499.

Jenkins ML, Kirk MA: *Characterization of radiation damage by transmission electron microscopy*, 2001, Inst. of Physics Publishing.

Minor AM, Stach EA, Morris JW Jr.: *In situ* nanoindentation in a transmission electron microscope. In Nabarro FRN, Hirth JP, editors: *Dislocations in solids*, vol 13, 2007, North-Holland, p. 453.

Williams DB, Carter CB: *Transmission electron microscopy: a textbook for materials science*, 2009, Springer.

Computer Simulation

Bacon DJ, Osetsky YN, Rodney D: Dislocation−obstacle interactions at the atomic level. In Hirth JP, Kubin L, editors: *Dislocations in solids*, vol 15, 2009, North-Holland, p. 1.

Bulatov VV, Cai W: *Computer simulations of dislocations*, 2006, Oxford University Press.

Cherkaoui M, Capolungo L: *Atomistic and continuum modeling of nanocrystalline materials: deformation mechanisms and scale transition*, 2009, Springer.

Finnis M: *Interatomic forces in condensed matter*, 2003, Oxford University Press.

Frederiksen SL, Jacobsen K: Density functional theory studies of screw dislocation core structures in bcc metals, *Phil Mag* 83:365, 2003.

Guo ZX, editor: *Multiscale materials modelling: fundamentals and applications*, 2007, Woodhead Publishing.

Li J, Ngan AHW, Gumbsch P: Atomistic modeling of mechanical behavior, *Acta Mater* 51:5711, 2003.

Phillips R: *Crystals, defects and microstructures: modeling across scales*, 2001, Cambridge University Press.

Van Swygenhoven H, Derlet PM: Atomistic simulations of dislocations in FCC metallic nanocrystalline materials. In Hirth JP, editor: *Dislocations in solids*, vol 14, 2008, North-Holland, p. 1.

Zbib HM, Khraishi TA: Size effects and dislocation−wave interaction in dislocation dynamics. In Hirth JP, editor: *Dislocations in solids*, vol 14, 2008, North-Holland, p. 207.

CHAPTER 3

Movement of Dislocations

3.1 CONCEPT OF SLIP

There are two basic types of dislocation movement. *Glide* or *conservative motion* occurs when the *dislocation moves in the surface which contains both its line and Burgers vector*: a dislocation able to move in this way is *glissile*, one which cannot is *sessile*. *Climb* or *non-conservative motion* occurs when the *dislocation moves out of the glide surface, and thus normal to the Burgers vector*. Glide of many dislocations results in *slip*, which is the most common manifestation of plastic deformation in crystalline solids. It can be envisaged as sliding or successive displacement of one plane of atoms over another on so-called *slip planes*. Discrete blocks of crystal between two slip planes remain undistorted as illustrated in Fig. 3.1. Further deformation occurs either by more movement on existing slip planes or by the formation of new slip planes.

The slip planes and slip directions in a crystal have specific crystallographic forms. The slip planes are normally the planes with the highest density of atoms, i.e. those which are most widely spaced, and the direction of slip is the direction in the slip plane corresponding to one of the shortest lattice translation vectors. Often, this direction is one in which the atoms are most closely spaced. It was seen in section 1.2 that the shortest lattice vectors are $\frac{1}{2}\langle 111 \rangle, \frac{1}{2}\langle 110 \rangle$ and $\frac{1}{3}\langle 11\bar{2}0 \rangle$ in the body-centered cubic, face-centered cubic and close-packed hexagonal systems, respectively. Thus, in close-packed hexagonal crystals, slip often occurs on the (0001) basal plane in directions of the type $\langle 11\bar{2}0 \rangle$ and in face-centered cubic metals on {111} planes in $\langle 110 \rangle$ directions. In body-centered cubic metals the slip direction is the $\langle 111 \rangle$ close-packed direction, but the slip plane is not well defined on a macroscopic scale. Microscopic evidence suggests that slip occurs on {112} and {110} planes and that {110} slip is preferred at low temperatures. A slip plane and a slip direction in the plane constitute a *slip system*. Face-centered cubic crystals have four {111} planes with three $\langle 110 \rangle$ directions in each (see Fig. 1.7), and therefore have twelve {111} $\langle 110 \rangle$ slip systems.

Introduction to Dislocations.

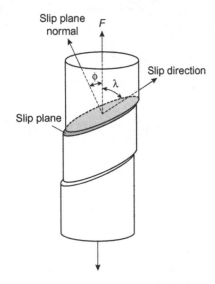

FIGURE 3.1

Illustration of the geometry of slip in a cylindrical crystal. Note that $(\phi + \lambda) \neq 90°$ in general.

Slip results in the formation of steps on the surface of the crystal. These are readily detected if the surface is carefully polished before plastic deformation. Figure 3.2 is an example of slip in a 3.25 per cent silicon iron crystal; the line diagram illustrates the appearance of a section through the crystal normal to the surface.

A characteristic shear stress is required for slip. Consider the crystal illustrated in Fig. 3.1 which is being deformed in tension by an applied force F along the cylindrical axis. If the cross-sectional area is A the tensile stress parallel to F is $\sigma = F/A$. The force has a component $F \cos \lambda$ in the slip direction, where λ is the angle between F and the slip direction. This force acts over the slip surface which has an area $A/\cos \phi$, where ϕ is the angle between F and the normal to the slip plane. Thus the *shear stress τ, resolved on the slip plane in the slip direction*, is

$$\tau = \frac{F}{A} \cos \phi \; \cos \lambda \qquad (3.1)$$

The symbol τ will be used throughout this book to denote the shear stress resolved in this way. If F_c is the tensile force required to start slip, the corresponding value of the shear stress τ_c is called the *critical resolved shear stress* (CRSS) *for slip*. It has been found in some crystals which deform on a single

(a)   (b)

FIGURE 3.2

(a) Straight slip bands on a single crystal of 3.25 per cent silicon iron. (*From Hull*, Proc. Roy. Soc. *A274, 5, 1963*.) (b) Sketch of a section across the slip bands normal to surface shown in (a). Each band is made up of a large number of slip steps on closely spaced parallel slip planes.

slip system that τ_c is independent of the orientation of the crystal. The quantity $\cos \phi \cos \lambda$ is known as the *Schmid factor*.

3.2 DISLOCATIONS AND SLIP

In Section 1.4 it was shown that the theoretical shear stress for slip is many times greater than the experimentally observed stress, i.e. τ_c. The low value of τ_c can be accounted for by the movement of dislocations. Consider the edge dislocation represented in Fig. 1.18. This could be formed in a different way to that described in Chapter 1, as follows: cut a slot along *AEFD* in the crystal shown in Fig. 3.3 and displace the top surface of the cut *AEFD* one lattice spacing over the bottom surface in the direction *AB*. An extra half-plane *EFGH* and a dislocation line *FE* are formed and the distortion produced is identical to that of Fig. 1.18. Although it is emphasized that dislocations are not formed in this way in practice, this approach demonstrates that the dislocation can be defined as the *boundary between the slipped and unslipped parts of the crystal*. Apart from the immediate region around the dislocation core *FE*, the atoms across *AEFD* are in perfect registry. The distortion due to the dislocations in Figs 1.18 and 3.3 has been described by giving all points on one side of an imagined cut – *ABCD* in Fig. 1.15 and *AEFD* in Fig. 3.3 – the same displacement relative to points on the other side; this displacement is the *Burgers vector* (section 1.4). These defects are *Volterra* dislocations, named after the Italian mathematician who first considered such distortions. More general forms arising from variable displacements are possible in principle, but are not treated here.

Only a relatively small applied stress is required to move the dislocation along the plane *ABCD* of the crystal in the way demonstrated in Fig. 3.4. This can be understood from the following argument. Well away from the dislocation, the atom spacings are close to the perfect crystal values, and a

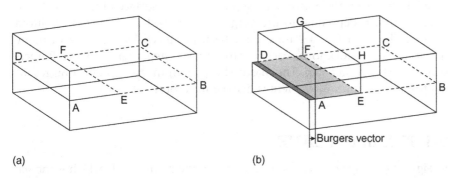

(a)   (b)

FIGURE 3.3
Formation of a pure edge dislocation *FE*.

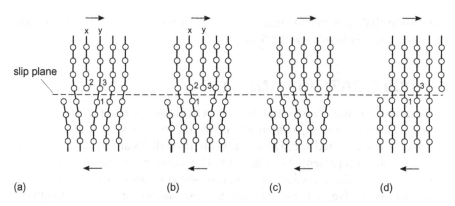

slip plane

(a)   (b)   (c)   (d)

FIGURE 3.4
Movement of an edge dislocation: the arrows indicate the applied shear stress.

shear stress as high as the theoretical value of equation (1.5) would be required to slide them all past each other at the same time. Near the dislocation line itself, some atom spacings are far from the ideal values, and small relative changes in position of only a few atoms are required for the dislocation to move. For example, a small shift of atom 1 relative to atoms 2 and 3 in Fig. 3.4(a) effectively moves the extra half-plane from x to y (Fig. 3.4(b)), and this process is repeated as the dislocation continues to glide (Figs 3.4(c), (d)). The applied stress required to overcome the lattice resistance to the movement of the dislocation is the *Peierls–Nabarro stress* (see Chapter 10) and is much smaller than the theoretical shear stress of a perfect crystal.

Figure 3.4 demonstrates why the Burgers vector is the most important parameter of a dislocation. Two neighboring atoms (say 1 and 3) on sites adjacent across the slip plane are displaced relative to each other by the Burgers vector when the dislocation glides past. Thus, *the slip direction* (see Fig. 3.1) *is necessarily always parallel to the Burgers vector of the dislocation responsible for slip.* The glide of one dislocation across the slip plane to the surface of the crystal produces a surface step equal to the Burgers vector. Each surface step produced by a slip band in Fig. 3.2 must have been produced by the glide of many thousands of dislocations, each contributing a step (equal to the Burgers vector) to the overall step. The plastic shear strain in the slip direction resulting from dislocation movement is derived in section 3.9.

3.3 THE SLIP PLANE

In Fig. 3.3 the edge dislocation has moved in the plane *ABCD* which is the *slip plane*. This is uniquely defined as *the plane which contains both the line and the Burgers vector of the dislocation.* The glide of an edge dislocation is limited, therefore, to a specific plane. The movement of a screw dislocation, for example from AA' to BB' in Fig. 3.5, can also be envisaged to take place in a slip plane, i.e. $LMNO$, and a slip step is formed. However, the line of the screw dislocation and the Burgers vector do not define a unique plane and the glide of the dislocation is not restricted to a specific plane. It will be noted that the displacement of atoms and hence the slip step associated with the movement of a screw dislocation is parallel to the dislocation line, for that is the direction of its Burgers vector. This can be demonstrated further by considering a plan view of the atoms above and below a slip plane containing a screw dislocation (Fig. 3.6). Movement of the screw dislocation produces a displacement b parallel to the dislocation line.

The direction in which a dislocation glides under stress can be determined by physical reasoning. Consider material under an applied shear stress (Fig. 3.7(a)) so that it deforms plastically by glide in the manner indicated in Fig. 3.7(b). From the description of section 3.2, a dislocation responsible for this deformation must have its Burgers vector in the direction shown. It is seen from Figs 3.7(c) and (d) that a positive edge dislocation glides to the left in order that the extra half-plane produces a step on the left-hand face as indicated, whereas a negative edge dislocation glides to the right. A right-handed screw glides towards the front in order to extend the surface step in the required manner (Fig. 3.7 (e)), whereas a left-handed screw glides towards the back (Fig. 3.7(f)). These observations demonstrate that (a) *dislocations of opposite sign glide in opposite directions under the same stress,* as expected of physical opposites, and (b) *for dislocation glide a shear stress must act on the slip plane in the direction of the Burgers vector,* irrespective of the direction of the dislocation line.

In the examples illustrated in Figs 3.3, 3.5 and 3.7 it has been assumed that the moving dislocations remain straight. However, dislocations are generally bent and irregular, particularly after plastic deformation, as can be seen in the electron micrographs presented throughout this book. A more general shape of a dislocation is shown in Fig. 3.8(a). The boundary separating the slipped and unslipped regions of the crystal is curved, i.e. the dislocation is curved, but the Burgers vector is the same all along its length. It follows that at point E the dislocation line is *normal* to the vector and is therefore *pure*

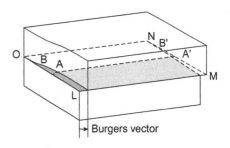

FIGURE 3.5
Formation and then movement of a pure screw dislocation AA' to BB' by slip.

FIGURE 3.6
Arrangement of atoms around a screw dislocation. Open circles above plane of diagram, filled circles below (for right-handed screw).

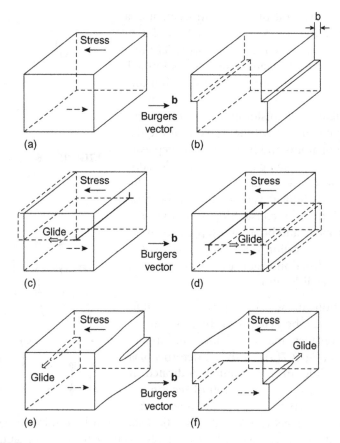

FIGURE 3.7

Plastic deformation by glide of edge and screw dislocations under the applied shear stress shown.

edge and at *S* is *parallel* to the vector and is *pure screw*. The remainder of the dislocation (*M*) has a *mixed edge and screw* character. The Burgers vector **b** of a mixed dislocation, *XY* in Fig. 3.8(b), can be resolved into two components by regarding the dislocation as two coincident dislocations; a pure edge with vector $\mathbf{b_1}$ of length $b \sin \theta$ at right angles to *XY*, and a pure screw with vector $\mathbf{b_2}$ of length $b \cos \theta$ parallel to *XY*:

$$\mathbf{b_1} + \mathbf{b_2} = \mathbf{b} \qquad (3.2)$$

3.4 CROSS SLIP

In general, screw dislocations tend to move in certain crystallographic planes (see, for example, dissociation of perfect dislocations, section 5.3). Thus, in

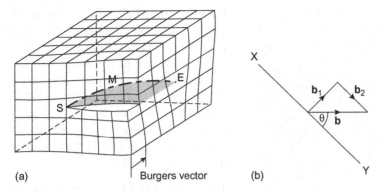

(a)   Burgers vector   (b)

FIGURE 3.8
Mixed dislocations. (a) The curved dislocation *SME* is pure edge at *E* and pure screw at *S*. (b) Burgers vector **b** of dislocation *XY* is resolved into a pure edge component $\mathbf{b}_1$ and a pure screw component $\mathbf{b}_2$.

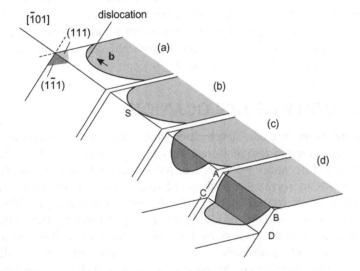

FIGURE 3.9
Sequence of events (a), (b), (c) in cross slip in a face-centered cubic metal. The [$1\bar{1}1$] direction is common to the (111) and ($1\bar{1}1$) close-packed planes. A screw dislocation at *S* is free to glide in either of these planes. Cross slip produces a non-planar slip surface. Double cross slip is shown in (d).

face-centered cubic metals the screw dislocations move in {111} type planes, but can switch from one {111} type plane to another if it contains the direction of **b**. This process, known as *cross slip*, is illustrated in Fig. 3.9. In Fig. 3.9 (a) a dislocation line, Burgers vector $\mathbf{b} = \frac{1}{2}[\bar{1}01]$, is gliding to the left in the (111) plane under the action of an applied shear stress. The only other {111} plane containing this slip vector is ($1\bar{1}1$). Suppose that as the loop expands

40 μm

FIGURE 3.10

Cross slip on the polished surface of a single crystal of 3.25 per cent silicon iron.

the local stress field which is producing dislocation motion changes so that motion is preferred on $(1\bar{1}1)$ instead of (111). Unlike edge and mixed dislocations, which have a unique glide plane, a pure screw segment is free to move in both $(1\bar{1}1)$ and (111) planes and so cross slip can occur at S (Fig. 3.9(b)). Glide of the dislocation then occurs on the $(1\bar{1}1)$ plane (Fig. 3.9(c)). *Double cross slip* is illustrated in Fig. 3.9(d).

The cross slip of moving dislocations is readily seen by transmission electron microscopy because a moving dislocation leaves a track which slowly fades. In body-centered cubic metals the slip plane is less well defined and the screw dislocation with $\mathbf{b} = \frac{1}{2}\langle 111 \rangle$ can glide on three $\{110\}$ planes and three $\{112\}$ planes. Slip often wanders from one plane to another producing wavy slip lines on prepolished surfaces. An example of the result of cross slip is shown in Fig. 3.10.

3.5 VELOCITY OF DISLOCATIONS

Dislocations move by glide at velocities which depend on the applied shear stress, purity of crystal, temperature and type of dislocation. A direct method of measuring dislocation velocity was developed by Johnston and Gilman using etch pits to reveal the position of dislocations at different stages of deformation, as illustrated in Fig. 2.13. A crystal containing freshly introduced dislocations, usually produced by lightly deforming the surface, is subjected to a constant stress pulse for a given time. From the positions of the dislocations before and after the stress pulse, the distance each dislocation has moved, and hence the average dislocation velocity, can be determined. By repeating the experiment for different times and stress levels the velocity can be determined as a function of stress as shown in Fig. 3.11(a) for lithium fluoride. The dislocation velocity was measured over 12 orders of magnitude and was a very sensitive function of the resolved shear stress. In the range of velocities between 10^{-9} and $10^{-3}\,\mathrm{ms}^{-1}$ the logarithm of the velocity varies linearly with the logarithm of the applied stress, thus

$$v \propto \left(\frac{\tau}{\tau_0}\right)^n \tag{3.3}$$

where v is velocity, τ is the applied shear stress resolved in the slip plane, τ_0 is the shear stress for $v = 1\,\mathrm{m\,s}^{-1}$, n is a constant and was found experimentally

FIGURE 3.11

(a) Stress dependence of the velocity of edge and screw dislocations in lithium fluoride. (*From Johnston and Gilman*, J. Appl. Phys. *30, 129, 1959.*) (b) Stress dependence of the velocity of edge dislocations in 3.25 per cent silicon iron at four temperatures. (*After Stein and Low*, J. Appl. Phys. *31, 362, 1960.*)

to be ~25 for lithium fluoride. It must be emphasized that equation (3.3) is purely empirical and implies no physical interpretation of the mechanism of dislocation motion.

The velocity of edge and screw components was measured independently and in the low velocity range edge dislocations moved 50 times faster than screw dislocations. There is a critical stress, which represents the onset of plastic deformation, required to start the dislocations moving. The effect of temperature on dislocation velocity is illustrated in Fig. 3.11(b) for results obtained by Stein and Low who studied 3.25 per cent silicon iron by the same method. The dislocation velocities were only measured below 10^{-4} m s^{-1} and

FIGURE 3.12

The dependence of dislocation velocity on applied shear stress. The data are for 20°C except Ge (450°C) and Si (850°C). *(After Haasen, Physical Metallurgy, Cambridge University Press.)*

therefore the curves do not show the bending over found in lithium fluoride. The curves are of the same form as equation (3.3). At 293 K, $n \sim 35$, and at 78 K, $n \sim 44$. τ_0 increased with decreasing temperature.

The stress dependence of dislocation velocity varies significantly from one material to another, as illustrated in Fig. 3.12. For a given material, the velocity of transverse shear-wave propagation is the limiting velocity for uniform dislocation motion. However, damping forces increasingly oppose motion when the velocity increases above about 10^{-3} of this limit, and thus n in equation (3.3) decreases rapidly in this range. Studies on face-centered cubic and hexagonal close-packed crystals have shown that at the critical resolved shear stress for macroscopic slip, dislocation velocity is approximately 1 m s^{-1} ($\sim 10^{-3}$ of shear wave velocity) and satisfies

$$v = A\tau^m \tag{3.4}$$

where A is a material constant, m is approximately 1 at 300 K in pure crystals, and increases to the range 2–5 with alloying and to 4–12 at 77 K. This form of stress-dependence is used in the dislocation dynamics computer simulations described in section 2.4.

When $m = 1$, equation (3.4) is usually written as

$$v = \frac{b}{B}\tau \tag{3.5}$$

where B is the *drag* (or *friction*) *coefficient*. In the velocity range where this applies (~ 0.1 to 1000 ms^{-1}), B is dominated by scattering of lattice vibrations (*phonons*). These are less numerous at low temperature and so B decreases. Effects due to phonon damping and the limitation of shear wave velocity have been studied by molecular dynamics computer simulation of dislocation glide (section 2.4). Velocity is plotted against τ for an edge dislocation gliding on a prism plane of zirconium (hexagonal close-packed) at three temperatures in Fig. 3.13(a) and on a {111} plane of nickel (face-centered cubic) with aluminum atoms in solution at 300 K in Fig. 3.13(b). B is generally found to be of the order of 10 μPa s in such simulations. Additional effects such as thermoelastic dissipation and radiative emission of phonons can be important in other materials.

(a)   Applied shear stress, MPa

(b)   Applied shear stress, MPa

FIGURE 3.13

Dislocation velocity versus applied shear stress obtained by computer simulation of an edge dislocation. (a) Pure zirconium at three temperatures. (*From Khater and Bacon, Acta Mater. 58 2978, 2010. Copyright (2010) with permission from Elsevier.* (b) Pure nickel and Ni-Al solid solutions at 300 K. (*With permission from Rodary, Rodney, Proville, Brechet and Martin, Phys. Rev. B 70, 05411, 2004. Copyright (2004) by the American Physical Society.*)

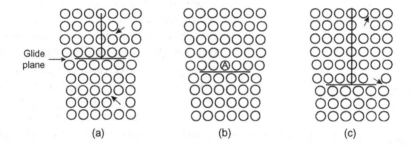

(a)   (b)   (c)

FIGURE 3.14

Positive and negative climb of an edge dislocation. Arrows indicate sense of vacancy motion. In (b) the dislocation is centered on the row of atoms *A* normal to the plane of the diagram. If the vacancies in the crystal diffuse to the dislocation at *A* the dislocation will climb in a positive sense as in (a). If vacancies are generated at the dislocation line and then diffuse away the dislocation will climb in the negative sense as in (c).

3.6 CLIMB

At low temperatures where diffusion is difficult, and in the absence of a non-equilibrium concentration of point defects, the movement of dislocations is restricted almost entirely to glide. However, at higher temperatures an edge dislocation can move out of its slip plane by a process called *climb*. Consider the diagram of an edge dislocation in Fig. 3.14. If the row of atoms *A* normal to the plane of the diagram is removed, the dislocation line moves up one atom spacing out of its original slip plane; this is called *positive climb*.

FIGURE 3.15

Displacement of a segment of line **l** by **s**. The area of the shaded element is $|\mathbf{l} \times \mathbf{s}|$.

FIGURE 3.16

A pair of jogs on an edge dislocation.

Similarly, if a row of atoms is introduced below the extra half-plane the dislocation line moves down one atom spacing, *negative climb*. Positive climb can occur by either diffusion of vacancies to *A* or the formation of a self-interstitial atom at *A* and its diffusion away. Negative climb can occur either by an interstitial atom diffusing to *A* or the formation of a vacancy at *A* and its diffusion away.

More generally, if a small segment l of line undergoes a small displacement **s** (Fig. 3.15), the local change in volume is

$$\mathrm{d}V = \mathbf{b}\cdot\mathbf{l}\times\mathbf{s} = \mathbf{b}\times\mathbf{l}\cdot\mathbf{s} \qquad (3.6)$$

for during the movement the two sides of the area element $\mathbf{l}\times\mathbf{s}$ are displaced by **b** relative to each other. The glide plane of the element is by definition perpendicular to $\mathbf{b}\times\mathbf{l}$, and so dV is zero when either s is perpendicular to $\mathbf{b}\times\mathbf{l}$ or $\mathbf{b}\times\mathbf{l} = 0$, which means the element is pure screw. In both cases **s** lies in the glide plane and this is the condition for glide (conservative motion) discussed previously. For other cases, volume is not conserved (dV $\neq$ 0) and the motion is climb, the number of point defects required being dV/Ω, where Ω is the volume per atom. The mass transport involved occurs by diffusion and therefore climb requires thermal activation. The most common climb processes involve the diffusion of vacancies either towards or away from the dislocation.

It has been implied above that a complete row of atoms is removed simultaneously, whereas in practice individual vacancies or small clusters of vacancies diffuse to the dislocation. The effect of this is illustrated in Fig. 3.16 which shows climb of a short section of a dislocation line resulting in the formation of two steps called *jogs*. Both positive and negative climb proceeds by the nucleation and motion of jogs. Conversely, jogs are *sources* and *sinks* for vacancies.

Jogs are steps on the dislocation which move it from one atomic slip plane to another. Steps which displace it on the same slip plane are called *kinks*. The two are distinguished in Fig. 3.17. Jogs and kinks are short elements of dislocation with the same Burgers vector as the line on which they lie, and the usual rules apply for their conservative and non-conservative movement. Thus, the kink, having the same slip plane as the dislocation line, does not impede glide of the line. (In fact, it may assist it – section 10.3.) Similarly, the jog on an edge dislocation (Fig. 3.17(c)) does not affect glide. The jog on a screw dislocation (Fig. 3.17(d)) has edge character, however, and can only glide along the line; movement at right angles to the Burgers vector requires climb. This impedes glide of the screw and results in point defect production during slip (Chapter 7). Since E_f^i is much larger than E_f^v in metals (section 1.3), it is very difficult for interstitial-producing jogs to move perpendicular to the line and vacancy production predominates.

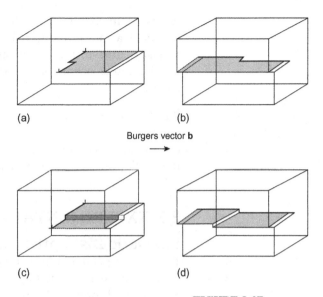

Burgers vector **b**

The jogs described have a height of one lattice spacing and a characteristic energy $E_j \sim 1$ eV (0.16 aJ) resulting from the increase in dislocation line length (section 7.2). They are produced extensively during plastic deformation by the intersection of dislocations (see Chapter 7), and exist even in well-annealed crystals, for there is a thermodynamic equilibrium number of jogs (*thermal jogs*) per unit length of dislocation given by

$$n_j = n_0 exp(-E_j/kT) \qquad (3.7)$$

where n_0, is the number of atom sites per unit length of dislocation (compare with equation (1.3)). n_j/n_0 is of the order of 10^{-5} at $T = 1000$ K and 10^{-17} at $T = 300$ K.

FIGURE 3.17

(a), (b) Kinks in edge and screw dislocations. (c), (d) Jogs in edge and screw dislocations. The slip planes are shaded.

The climb of dislocations by jog formation and migration is analogous to crystal growth by surface step transport. There are two possible mechanisms. In one, a pre-existing single jog (as in Fig. 3.17(c)), or a jog formed at a site without the need for thermal nucleation, migrates along the line by vacancy emission or absorption. This involves no change in dislocation length and the activation energy for jog diffusion is

$$(E_f^v + E_m^v) = E_d \qquad (3.8)$$

where E_m^v is the vacancy migration energy and E_d the activation energy for self-diffusion. In the other, thermal jogs are nucleated on an otherwise straight line and the jog migration energy is $(E_d + E_j)$. In most situations the first process dominates. The effective activation energy in some circumstances can be less than one half the value of E_d measured in the bulk, for the crystal is distorted at atom sites close to the dislocation line itself and E_d is smaller there. The process of mass transport occurring along the line, rather than to or from it, is known as *pipe diffusion*, and an example is discussed in section 3.8. Although jog diffusion in thermal equilibrium occurs at a rate proportional to $\exp(-E_d/kT)$, or a rate with E_d modified as just discussed, the actual rate of climb of the dislocation depends also on the forces acting on it. Several factors are involved, as discussed in section 4.7.

Pure screw dislocations have no extra half-plane and in principle cannot climb. However, a small edge component or a jog on a screw dislocation can provide a site for the start of climb. Two examples will serve to illustrate the climb process in both edge and predominantly screw dislocations.

3.7 EXPERIMENTAL OBSERVATION OF CLIMB

Prismatic Dislocation Loops

The shaded plane bounded by the curved line *SME* in Fig. 3.8 contains the Burgers vector **b** and is therefore the slip plane of the dislocation (section 3.3). If a dislocation forms a loop in a plane and **b** lies in that plane (Fig. 3.18(a)), the plane is the slip plane and the loop can expand or shrink by glide, depending on the forces acting on it (section 4.5). When the Burgers vector is not in the plane of the loop, the glide surface defined by the dislocation line and its Burgers vector is a cylindrical surface (Fig. 3.18 (b)). The dislocation is called a *prismatic dislocation*. It follows that the dislocation can only move conservatively, i.e. by glide, along the cylindrical surface and if the loop expands or shrinks *climb* must be occurring.

Numerous examples of prismatic loops have been observed using transmission electron microscopy. They can be formed in the following way. The supersaturation (excess concentration) of vacancies resulting from either rapid quenching from a high temperature (see section 1.3) or atomic displacements produced by irradiation with energetic particles may precipitate out in the form of a disc on a close-packed plane. If the disc is large enough, it is energetically favorable for it to collapse to produce a dislocation loop (Fig. 3.19). The Burgers vector of the loop is normal to the plane of the loop, so that an edge dislocation has been formed. In the presence of an excess concentration of vacancies the loop will expand by positive climb. Alternatively, at high temperature when the equilibrium concentration of vacancies in the crystal is

FIGURE 3.18

Glide plane and glide cylinder for dislocation loops with Burgers vector (a) in the loop plane and (b) inclined to the loop plane.

Glide cylinder or prism

Glide plane

(a)   (b)

(a)   (b)   (c)   (d)

FIGURE 3.19

Formation of a prismatic dislocation loop. (a) Represents a crystal with a large non-equilibrium concentration of vacancies. In (b) the vacancies have collected on a close-packed plane and in (c) the disc has collapsed to form an edge dislocation loop. (d) Loop formed by a platelet of self-interstitial atoms.

large (section 1.3) or if there is a nearby sink for vacancies, the loop will emit vacancies and shrink by negative climb. Figure 3.20 shows an example of the latter. The dislocation loops were formed in a thin sheet of aluminum by quenching. The sheet was thinned to about 100 nm and examined by transmission electron microscopy. The surfaces of the sheet are very effective sinks for vacancies and when the foil was heated in the microscope to allow thermal activation to assist in the formation and diffusion of vacancies, the stacking-fault and line-tension forces (section 4.7) acting on the dislocations induced the loops to shrink and disappear.

FIGURE 3.20

Electron transmission photographic sequence showing the shrinkage of $\frac{1}{3}\langle 111\rangle$ stacking fault dislocation loops in aluminum by negative climb. The foil was annealed at 102°C for 0, 213, 793 and 1301 min, respectively. *(From Tartour and Washburn*, Phil. Mag. *18, 1257, 1968.)*

In a similar fashion, platelets of self-interstitial atoms can form dislocations in irradiated materials (Fig. 3.19(d)) and can grow by absorption of self-interstitials created by radiation damage.

Helical Dislocations

Dislocations in the form of a long spiral have been observed in crystals which have been thermally treated to produce climb conditions. Thus, in Fig. 3.21 the helical dislocation in CaF_2 was formed by heating the crystal to a high temperature. A mechanism for the formation of helical dislocations is as follows. The dislocation *AB*, in Fig. 3.22, is pinned or locked at *A* and *B*, and is partly edge and partly screw in character. Motion of the dislocation in the plane *ABA'* corresponds to glide since this plane contains the line and the Burgers vector. Motion at right angles to this plane corresponds to climb.

FIGURE 3.21

Spiral or helical dislocation in CaF$_2$ (fluorite) revealed by the decoration technique. *(After Bontinck and Amelinckx*, Phil. Mag. *2, 94, 1957.)*

An excess of vacancies at a suitable temperature will cause the dislocation to climb. The configuration of the dislocation after a certain amount of climb is shown in Fig. 3.22(b). The dislocation now lies on the surface of a cylinder whose axis is parallel to the Burgers vector (i.e. prismatic dislocation) and it can glide on the surface of the cylinder. Further climb displaces each part of the dislocation in a direction normal to the surface of the cylinder. The dislocation will change into a double spiral by spiraling around node points A and B, as shown in Fig. 3.22(c). The radius of the spiral is smallest at the nodes, since, for a given number of vacancies, the angle of rotation is greater there. If A'B is small compared with AA', combination of prismatic glide and climb will result in the formation of a uniformly spaced spiral. The Burgers vector of the helical dislocations in Figs 3.21 and 3.22 must therefore lie along the axis of the helix.

The helix consists essentially of a screw dislocation parallel to the axis of the helix and a set of prismatic loops. This interpretation is made more apparent by considering the interaction between a screw dislocation and a dislocation loop with the same Burgers vector. Figure 3.23 illustrates how the dislocation acquires a helical turn by absorption of the loop. The direction of the Burgers vector and positive line

FIGURE 3.22

Formation of a helical dislocation by climb of a straight dislocation with a screw component. (a) A straight dislocation line *AB* and its projection in the plane *PN* which is normal to the Burgers vector **b** and passes through *B*. *A'* is the projection of *A* on to plane *PN*. (b) Change produced in the dislocation *AB* by climb. *AB* is now curved and lies on a cylinder whose axis is parallel to **b**. The dislocation can glide on this cylinder. The area *F* is proportional to the amount of material added or lost in climb. (c) Helical dislocation produced after further climb. The projection of this dislocation on to *PN* is the double spiral shown in the diagram. *(After Amelinckx, Bontinck, Dekeyser and Seitz*, Phil. Mag. *2, 355, 1957.)*

sense of the dislocation and loop are indicated by arrows. Application of the RH/FS convention (section 1.4) shows that the dislocation is a right-handed screw and the loop has interstitial nature in (a) and vacancy nature in (b). The sense of the resulting helical turn is different in (a) and (b). It would be reversed for a left-handed screw.

Loop + Screw dislocation = Helical turn

FIGURE 3.23
Schematic illustration of a right-handed screw dislocation acquiring a helical turn by reaction with a dislocation loop with the same Burgers vector: (a) interstitial loop; (b) vacancy loop.

3.8 CONSERVATIVE CLIMB

From the description of the basic types of dislocation movement at the beginning of this chapter, the heading of this section appears to be a contradiction. However, if a prismatic loop (Fig. 3.19) moves in its plane without shrinking or expanding at temperatures too low for bulk diffusion of point defects, it follows that climb is occurring without net loss or gain of vacancies or self-interstitial atoms. An example is seen in Fig. 3.24. Loop translation occurs by the transfer of vacancies around the loop by *pipe diffusion*, the vacancies producing positive climb at one side of the loop and negative climb at the other side. The process is called *conservative climb*.

3.9 PLASTIC STRAIN DUE TO DISLOCATION MOVEMENT

It is implicit in the description of dislocation behavior presented in preceding sections that dislocation movement, usually under the influence of an externally-applied load, results in *plastic strain*. This is in addition to the *elastic strain*, which is simply related to the external stress by Hooke's law (Chapter 4). The relation between plastic strain and the applied stress is more complicated and depends on factors such as temperature, applied strain rate and, in particular, the microstructure of the material: these aspects are discussed further in Chapter 10. There is, however, a simple relationship (the *Orowan equation*) between the plastic strain and the dislocation density

FIGURE 3.24
Sequence of electron micrographs showing the conservative climb motion of a dislocation loop, with a Burgers vector normal to the plane of the loop, due to its interaction with an edge dislocation. *(From Kroupa and Price, Phil Mag. 6, 243, 1961.)*

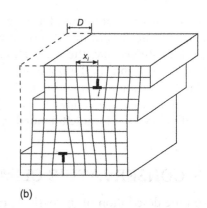

FIGURE 3.25

(a) Edge dislocations in a crystal subjected to an external shear stress resolved for slip. (b) Plastic displacement D produced by glide of the dislocations. Dislocation i has moved a distance x_i as shown.

defined as in section 1.4. It is based on the fact that when a dislocation moves, two atoms on sites adjacent across the plane of motion are displaced relative to each other by the Burgers vector **b** (section 3.2).

The relationship for *slip* is derived first. Consider a crystal of volume hld containing for simplicity straight edge dislocations (Fig. 3.25(a)). Under a high enough applied shear stress acting on the slip plane in the direction of **b**, as shown, the dislocations will glide, positive ones to the right, negative ones to the left. The top surface of the sample is therefore displaced plastically by D relative to the bottom surface as demonstrated in Fig. 3.25(b). If a dislocation moves completely across the slip plane through the distance d, it contributes b to the total displacement D. Since b is small in comparison to d and h, the contribution made by a dislocation which moves a distance x_i may be taken as the fraction (x_i/d) of b. Thus, if the number of dislocations which move is N, the total displacement is

$$D = \frac{b}{d} \sum_{i=1}^{N} x_i \tag{3.9}$$

and the macroscopic *plastic* shear strain ε is given by

$$\varepsilon = \frac{D}{h} = \frac{b}{hd} \sum_{i=1}^{N} x_i \tag{3.10}$$

This can be simplified by noting that the *average* distance moved by a dislocation $\bar{x}$ is

$$\bar{x} = \frac{1}{N} \sum_{i=1}^{N} x_i \qquad (3.11)$$

and that the density of mobile dislocations ρ_m is (Nl/hld). Hence

$$\varepsilon = b\rho_m \bar{x} \qquad (3.12)$$

The *strain rate* is therefore

$$\dot{\varepsilon} = \frac{d\varepsilon}{dt} = b\rho_m \bar{v} \qquad (3.13)$$

where $\bar{v}$ is the average dislocation velocity. The same relationships hold for screw and mixed dislocations. Also, since the average area of slip plane swept by a dislocation $\bar{A}$ equals $l\bar{x}$, an alternative to equation (3.12) is

$$\varepsilon = bn\bar{A} \qquad (3.14)$$

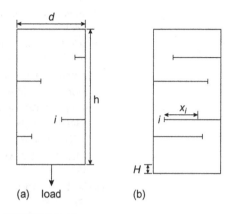

FIGURE 3.26

(a) Edge dislocations, denoted by their extra half planes, in a crystal under a tensile load. (b) The dislocations have climbed, producing a plastic elongation H.

where n is the number of lines per unit volume. It is emphasized that ρ_m appearing in the above equations is the *mobile* dislocation density, for dislocations which do not move do not contribute to the plastic strain.

Climb under an external tensile load is shown schematically in Fig. 3.26. When an edge dislocation climbs, an extra plane of thickness b is inserted into, or removed from, the crystal in the area over which the line moves. As in the analysis for slip, if a dislocation moves through distance x_i it contributes $b(x_i/d)$ to the plastic displacement H of the external surface. It is therefore easy to show that the total plastic *tensile* strain (H/h) parallel to the Burgers vector and the strain rate are given by equations (3.12) and (3.13) respectively. The same relations also hold for climb of mixed dislocations, except that b is then the magnitude of the edge component of the Burgers vector.

FURTHER READING

The books and paper listed under 'Dislocations' at the end of Chapter 1 provide excellent sources of further information.

Balluffi RW, Granato AV: Dislocations, vacancies and interstitials. In Nabarro FRN, editor: *Dislocations in solids*, vol 4, 1979, North-Holland, p. 1.

Cai W, Bulatov V, Chang V, Li JJ, Yip S: Dislocation core effects on mobility. In Nabarro FRN, Hirth JP, editors: *Dislocations in solids*, vol 12, 2004, North-Holland, p. 1.

Gilman JJ: *Micromechanics of flow in solids*, 1969, McGraw-Hill.

Gillis PP, Gilman JJ, Taylor JW: Stress dependences of dislocation velocities, *Phil Mag* 20:279, 1969.

Imura T: Dynamic studies of plastic deformation by means of high voltage electron microscopy, *Electron microscopy and strength of materials*, 1972, University of California Press, p. 104.

Indenbom VL, Lothe J, editors: *Elastic strain fields and dislocation mobility*, 1992, Elsevier.

Johnston WG: Yield points and delay times in single crystals, *J Appl Phys* 33:2716, 1962.

Johnston WG, Gilman JJ: Dislocation velocities, dislocation densities and plastic flow in lithium fluoride crystals, *J Appl Phys* 30:129, 1959.

Rosenfield AR, Hahn GT, Bement AL, Jaffee RI, editors: *Dislocation dynamics*, 1968, McGraw-Hill.

Stein DF, Low JW: Mobility of edge dislocations in silicon iron crystals, *J Appl Phys* 31:362, 1960.

Vreeland T Jr.: Dislocation velocity measurements. In Bunshaw RF, editor: *Techniques of metals research*, vol 2, 1968, Wiley, p. 341.

CHAPTER 4

Elastic Properties of Dislocations

4.1 INTRODUCTION

The atoms in a crystal containing a dislocation are displaced from their perfect lattice sites, and the resulting distortion produces a stress field in the crystal around the dislocation. The dislocation is therefore a source of *internal stress* in the crystal. For example, consider the edge dislocation in Fig. 1.18 (b). The region above the slip plane contains the extra half-plane forced between the normal lattice planes, and is in compression: the region below is in tension. The stresses and strains in the bulk of the crystal are sufficiently small for conventional elasticity theory to be applied to obtain them. This approach only ceases to be valid at positions very close to the center of the dislocation. Although most crystalline solids are elastically *anisotropic*, i.e. their elastic properties are different in different crystallographic directions, it is much simpler to use *isotropic elasticity* theory. This still results in a good approximation in most cases. From a knowledge of the elastic field, the energy of the dislocation, the force it exerts on other dislocations, its energy of interaction with point defects, and other important characteristics, can be obtained. The elastic field produced by a dislocation is not affected by the application of stress from *external* sources: the total stress on an element within the body is the superposition of the internal and external stresses.

4.2 ELEMENTS OF ELASTICITY THEORY

The *displacement* of a point in a strained body from its position in the unstrained state is represented by the vector

$$\mathbf{u} = [u_x, u_y, u_z] \qquad (4.1)$$

The components u_x, u_y, u_z represent projections of $\mathbf{u}$ on the x, y, z axes, as shown in Fig. 4.1. In *linear* elasticity, the nine components of *strain* are

63

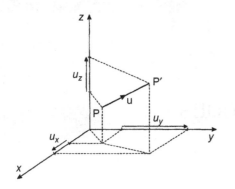

FIGURE 4.1

Displacement of P to P' by displacement vector **u**.

given in terms of the first derivatives of the displacement components thus:

$$e_{xx} = \frac{\partial u_x}{\partial x}$$

$$e_{yy} = \frac{\partial u_y}{\partial y} \qquad (4.2)$$

$$e_{zz} = \frac{\partial u_z}{\partial z}$$

and

$$e_{yz} = e_{zy} = \frac{1}{2}\left(\frac{\partial u_y}{\partial z} + \frac{\partial u_z}{\partial y}\right)$$

$$e_{zx} = e_{xz} = \frac{1}{2}\left(\frac{\partial u_z}{\partial x} + \frac{\partial u_x}{\partial z}\right) \qquad (4.3)$$

$$e_{xy} = e_{yx} = \frac{1}{2}\left(\frac{\partial u_x}{\partial y} + \frac{\partial u_y}{\partial x}\right)$$

The magnitude of these components is $\ll 1$. Partial differentials are used because in general each displacement component is a function of position (x, y, z). The three strains defined in (4.2) are the *normal* strains. They represent the fractional change in length of elements parallel to the x, y and z axes respectively, e.g. the length l_x of an element in the x direction is changed to $l_x (1 + e_{xx})$. The six components defined in (4.3) are the *shear* strains, and they also have simple physical meaning. This is demonstrated by e_{xy} in Fig. 4.2(a), in which a small area element $ABCD$ in the xy plane has been strained to the shape $AB'C'D'$ without change of area. The angle between the

(a)

(b)

FIGURE 4.2

(a) Pure shear and (b) simple shear of an area element in the xy plane.

sides AB and AD initially parallel to x and y respectively has decreased by $2e_{xy}$. By rotating, but not deforming, the element as in Fig. 4.2(b), it is seen that the element has undergone a simple shear. The simple shear strain often used in engineering practice is $2e_{xy}$, as indicated.

The volume V of a small volume element is changed by strain to $(V + \Delta V) = V(1 + e_{xx})(1 + e_{yy})(1 + e_{zz})$. The fractional change in volume Δ, known as the *dilatation*, is therefore

$$\Delta = \Delta V/V = (e_{xx} + e_{yy} + e_{zz}) \tag{4.4}$$

Δ is independent of the orientation of the axes x, y, z.

In elasticity theory, an element of volume experiences forces via stresses applied to its surface by the surrounding material. *Stress* is the force per unit area of surface. A complete description of the stresses acting therefore requires not only specification of the magnitude and direction of the force but also of the orientation of the surface, for as the orientation changes so, in general, does the force. Consequently, nine components must be defined to specify the state of stress. They are shown with reference to an elemental cube aligned with the x, y, z axes in Fig. 4.3(a). The component σ_{ij}, where i and j can be x, y or z, is defined as the force per unit area exerted in the $+i$ direction on a face with outward normal in the $+j$ direction by the material *outside* upon the material *inside*. For a face with outward normal in the $-j$ direction, i.e. the bottom and back faces shown in Fig. 4.3(b), σ_{ij} is the force per unit area exerted in the $-i$ direction. For example, σ_{yz} acts in the positive y direction on the top face and the negative y direction on the bottom face.

The six components with $i \neq j$ are the *shear* stresses. (As explained in section 3.1, it is customary in dislocation studies to represent the shear stress acting

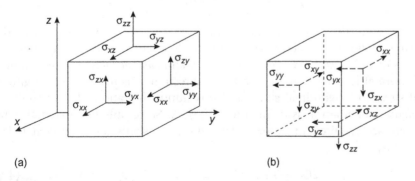

(a)   (b)

FIGURE 4.3
Components of stress acting on (a) the top and front faces and (b) the bottom and back faces of an elemental cube.

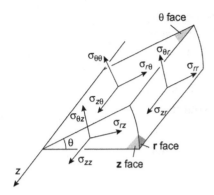

FIGURE 4.4

Components of stress in cylindrical polar coordinates.

on the slip plane in the slip direction of a crystal by the symbol τ.) By considering moments of forces taken about x, y and z axes placed through the center of the cube, it can be shown that rotational equilibrium of the element, i.e. net couple = 0, requires

$$\sigma_{yz} = \sigma_{zy} \quad \sigma_{zx} = \sigma_{xz} \quad \sigma_{xy} = \sigma_{yx} \tag{4.5}$$

Thus, the order in which subscripts i and j is written is immaterial. The three remaining components σ_{xx}, σ_{yy}, σ_{zz} are the *normal* components. From the definition given above, a *positive* normal stress results in *tension* and a *negative* one in *compression*. The effective pressure acting on a volume element is therefore

$$p = -\frac{1}{3}(\sigma_{xx} + \sigma_{yy} + \sigma_{zz}) \tag{4.6}$$

For some problems, it is more convenient to use cylindrical polar coordinates (r, θ, z). The stresses are still defined as above, and are shown in Fig. 4.4. The notation is easier to follow if the second subscript j is considered as referring to the face of the element having a constant value of the coordinate j.

The relationship between stress and strain in linear elasticity is *Hooke's Law*, in which each stress component is linearly proportional to each strain. For isotropic solids, only two proportionality constants are required:

$$\begin{aligned}
\sigma_{xx} &= 2Ge_{xx} + \lambda(e_{xx} + e_{yy} + e_{zz}) \\
\sigma_{yy} &= 2Ge_{yy} + \lambda(e_{xx} + e_{yy} + e_{zz}) \\
\sigma_{zz} &= 2Ge_{xx} + \lambda(e_{xx} + e_{yy} + e_{zz}) \\
\sigma_{xy} &= 2Ge_{xy} \quad \sigma_{yz} = 2Ge_{yz} \quad \sigma_{zx} = 2Ge_{zx}
\end{aligned} \tag{4.7}$$

λ and G are the Lamé constants, but G is more commonly known as the *shear modulus*. Other elastic constants are frequently used, the most useful being *Young's modulus, E, Poisson's ratio, ν,* and *bulk modulus, K*. Under uniaxial, normal loading in the longitudinal direction, E is the ratio of longitudinal stress to longitudinal strain and ν is minus the ratio of lateral strain to longitudinal strain. K is defined to be $-p/\Delta$. Since only two material parameters are required in Hooke's law, these constants are interrelated. For example,

$$E = 2G(1 + \nu) \quad \nu = \lambda/2(\lambda + G) \quad K = E/3(1 - 2\nu) \tag{4.8}$$

Typical values of E and ν for metallic and ceramic solids lie in the ranges $40-600$ GNm^{-2} and $0.2-0.45$ respectively.

The internal energy of a body is increased by strain. The *strain energy* per unit volume is one-half the product of stress times strain for each component. Thus, for an element of volume dV, the elastic strain energy is

$$dE_{el} = \frac{1}{2}dV \sum_{i=x,y,z} \sum_{j=x,y,z} \sigma_{ij}e_{ij} \qquad (4.9)$$

and similarly for polar coordinates.

4.3 STRESS FIELD OF A STRAIGHT DISLOCATION

Screw Dislocation

The elastic distortion around an infinitely-long, straight dislocation can be represented in terms of a cylinder of elastic material. Consider the screw dislocation AB shown in Fig. 4.5(a); the elastic cylinder in Fig. 4.5(b) has been deformed to produce a similar distortion. A radial slit $LMNO$ was cut in the cylinder parallel to the z-axis and the cut surfaces displaced rigidly with respect to each other by the distance b, the magnitude of the Burgers vector of the screw dislocation, in the z-direction.

The elastic field in the dislocated cylinder can be found by direct inspection. First, it is noted that there are no displacements in the x and y directions:

$$u_x = u_y = 0 \qquad (4.10)$$

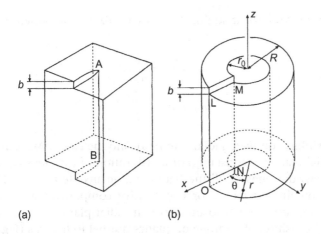

(a)   (b)

FIGURE 4.5

(a) Screw dislocation AB formed in a crystal. (b) Elastic distortion of a cylindrical tube simulating the distortion produced by the screw dislocation in (a).

Secondly, the displacement in the z-direction increases uniformly from zero to b as θ increases from 0 to 2π:

$$u_z = \frac{b\theta}{2\pi} = \frac{b}{2\pi}\tan^{-1}(y/x) \qquad (4.11)$$

It is then readily found from equations (4.2) and (4.3) that

$$e_{xx} = e_{yy} = e_{zz} = e_{xy} = e_{yx} = 0$$
$$e_{xz} = e_{zx} = -\frac{b}{4\pi}\frac{y}{(x^2+y^2)} = -\frac{b}{4\pi}\frac{\sin\theta}{r}$$
$$e_{yz} = e_{zy} = \frac{b}{4\pi}\frac{x}{(x^2+y^2)} = \frac{b}{4\pi}\frac{\cos\theta}{r} \qquad (4.12)$$

From equations (4.7) and (4.12), the components of stress are

$$\sigma_{xx} = \sigma_{yy} = \sigma_{zz} = \sigma_{xy} = \sigma_{yx} = 0$$
$$\sigma_{xz} = \sigma_{zx} = -\frac{Gb}{2\pi}\frac{y}{(x^2+y^2)} = -\frac{Gb\sin\theta}{2\pi}\frac{}{r}$$
$$\sigma_{yz} = \sigma_{zy} = \frac{Gb}{2\pi}\frac{x}{(x^2+y^2)} = \frac{Gb\cos\theta}{2\pi}\frac{}{r} \qquad (4.13)$$

The components in cylindrical polar coordinates (Fig. 4.4) take a simpler form. Using the relations

$$\sigma_{rz} = \sigma_{xz}\cos\theta + \sigma_{yz}\sin\theta$$
$$\sigma_{\theta z} = -\sigma_{xz}\sin\theta + \sigma_{yz}\cos\theta \qquad (4.14)$$

and similarly for the shear strains, the only non-zero components are found to be

$$e_{\theta z} = e_{z\theta} = \frac{b}{4\pi r}$$
$$\sigma_{\theta z} = \sigma_{z\theta} = \frac{Gb}{2\pi r} \qquad (4.15)$$

The field exhibits complete radial symmetry and the cut $LMNO$ can be made on any radial plane $\theta = $ constant. For a dislocation of *opposite* sign, i.e. a left-handed screw, the signs of all the field components are *reversed*. The elastic distortion contains no tensile or compressive components and consists of pure shear: $\sigma_{z\theta}$ acts parallel to the z-axis in radial planes of constant θ and $\sigma_{\theta z}$ acts in the fashion of a torque on planes normal to the axis (Fig. 4.4).

For a crystal of finite length the $\sigma_{\theta z}$ torque results in a twist about the cylinder axis (*Eshelby twist*). An example is provided by growth of PbSe nanowires

(a)   (b)   (c)

FIGURE 4.6

(a) Scanning electron micrograph of a branched nanowire with a twist (inset: high-resolution image). Sketch of a wire containing an axial screw dislocation (b) without and (c) with branch wires. The end of the main wire is free to rotate and contains the surface step that mediates growth, *(From Zhu, Peng, Marshall, Barnett, Nix and Cui, Nature Nanotechnology **3**, 477, 2008. Reprinted with permission from Macmillan Publishers Ltd: copyright (2008).)*

from vapor (Fig. 4.6). The main wire (50 μm long) has a twist, as revealed by the orientation of shorter branched nanowires that have grown perpendicular to it (Fig. 4.6(a)). Growth of the main wire was mediated by the surface step due to an axial screw dislocation (see section 2.3 and Fig. 2.10(c)), which created the twist (Figs 4.6(b) and (c)).

The stresses and strains are proportional to $1/r$ and therefore diverge to infinity as $r \to 0$. Solids cannot withstand infinite stresses, and for this reason the cylinder in Fig. 4.5 is shown as hollow with a hole of radius r_0. Real crystals are not hollow, of course, and so as the center of a dislocation in a crystal is approached, elasticity theory ceases to be valid and a non-linear, atomistic model must be used (see section 10.3). The region within which the linear-elastic solution breaks down is called the *core* of the dislocation. From equation (4.15) it is seen that the stress reaches the theoretical limit (equation (1.5)) and the strain exceeds about 10% when $r \approx b$. A reasonable value for the *dislocation core radius* r_0 therefore lies in the range b to $4b$, i.e. $r_0 \lesssim 1$ nm in most cases.

Edge Dislocation

The stress field is more complex than that of a screw but can be represented in an isotropic cylinder in a similar way. Considering the edge dislocation in Fig. 4.7(a), the same elastic strain field can be produced in the cylinder by a rigid displacement of the faces of the slit by a distance b in the x-direction (Fig. 4.7(b)). The displacement and strains in the z-direction are zero and

(a)   (b)

FIGURE 4.7

(a) Edge dislocation formed in a crystal. (b) Elastic distortion of a cylindrical ring simulating the distortion produced by the edge dislocation in (a).

the deformation is called plane strain. Derivation of the field components is beyond the scope of the present treatment, however. The stresses are found to be

$$\sigma_{xx} = -Dy\frac{(3x^2 + y^2)}{(x^2 + y^2)^2}$$

$$\sigma_{yy} = Dy\frac{(x^2 - y^2)}{(x^2 + y^2)^2}$$

$$\sigma_{xy} = \sigma_{yx} = Dx\frac{(x^2 - y^2)}{(x^2 + y^2)^2}$$

$$\sigma_{zz} = \nu(\sigma_{xx} + \sigma_{yy})$$

$$\sigma_{xz} = \sigma_{zx} = \sigma_{yz} = \sigma_{zy} = 0 \qquad (4.16)$$

where

$$D = \frac{Gb}{2\pi(1 - \nu)}$$

The stress field has both dilational and shear components. The largest normal stress is σ_{xx} which acts parallel to the slip vector. Since the slip plane can be defined as $y = 0$, the maximum compressive stress (σ_{xx} negative) acts immediately above the slip plane and the maximum tensile stress (σ_{xx} positive) acts immediately below the slip plane. The effective pressure (equation (4.6)) on a volume element is

$$p = \frac{2}{3}(1 + \nu)D\frac{y}{(x^2 + y^2)} \qquad (4.17)$$

It is compressive above the slip plane and tensile below. These observations are implied qualitatively by the type of distortion illustrated in Figs 1.18 and 4.7(a).

As in the case of the screw, the signs of the components are reversed for a dislocation of opposite sign, i.e. a negative edge dislocation with extra half-plane along the negative y-axis. Again the elastic solution has an inverse dependence on distance from the line axis and breaks down when x and y tend to zero. It is valid only outside a core of radius r_0.

The elastic field produced by a *mixed* dislocation (Fig. 3.8(b)) having *edge and screw character* is obtained from the above equations by adding the fields of the edge and screw constituents, which have Burgers vectors given by equation (3.2). The two sets are independent of each other in isotropic elasticity.

4.4 STRAIN ENERGY OF A DISLOCATION

The existence of distortion around a dislocation implies that a crystal containing a dislocation is not in its lowest energy state. The extra energy is the *strain energy*. The total strain energy may be divided into two parts

$$E_{total} = E_{core} + E_{elastic\ strain} \qquad (4.18)$$

The elastic part, stored outside the core, may be determined by integration of the energy of each small element of volume. This is a simple calculation for the screw dislocation, because from the symmetry the appropriate volume element is a cylindrical shell of radius r and thickness dr. From equation (4.9), the elastic energy stored in this volume *per unit length* of dislocation is

$$dE_{el}(\text{screw}) = \frac{1}{2} 2\pi r\ dr(\sigma_{\theta z} e_{\theta z} + \sigma_{z\theta} e_{z\theta}) = 4\pi r\ dr G e_{\theta z}^2 \qquad (4.19)$$

Thus, from equation (4.15), the total elastic energy stored in the cylinder (Fig. 4.5) per unit length of dislocation is

$$E_{el}(\text{screw}) = \frac{Gb^2}{4\pi} \int_{r_0}^{R} \frac{dr}{r} = \frac{Gb^2}{4\pi} \ln\left(\frac{R}{r_0}\right) \qquad (4.20)$$

where R is the outer radius.

The above approach is much more complicated for other dislocations having less symmetric fields. It is generally easier to consider E_{el} as the work done in displacing the faces of the cut *LMNO* by b (Figs 4.5 and 4.7) against the resisting internal stresses. For an infinitesimal element of area dA of *LMNO*, the work done is

$$dE_{el}(\text{screw}) = \frac{1}{2}\sigma_{zy} b dA$$

$$dE_{el}(\text{edge}) = \frac{1}{2}\sigma_{xy} b dA \qquad (4.21)$$

with the stresses evaluated on $y = 0$. The factor $\frac{1}{2}$ enters because the stresses build up from zero to the final values given by equations (4.13) and (4.16)) during the displacement process. The element of area is a strip of width dx parallel to the z-axis, and so the total strain energy per unit length of dislocation is

$$E_{el}(\text{screw}) = \frac{Gb^2}{4\pi} \int_{r_0}^{R} \frac{dx}{x} = \frac{Gb^2}{4\pi} \ln\left(\frac{R}{r_0}\right)$$

$$E_{el}(\text{edge}) = \frac{Gb^2}{4\pi(1-\nu)} \int_{r_0}^{R} \frac{dx}{x} = \frac{Gb^2}{4\pi(1-\nu)} \ln\left(\frac{R}{r_0}\right) \qquad (4.22)$$

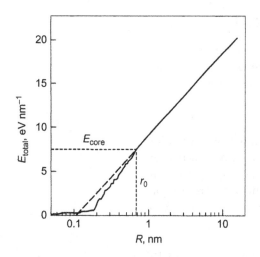

FIGURE 4.8

The strain energy within a cylinder of radius R that contains a straight edge dislocation along its axis. The data was obtained by computer simulation for a model of iron. *(Courtesy Yu. N. Osetsky.)*

The screw result is the same as equation (4.20). Strictly, equations (4.22) neglect small contributions from the work done on the core surface $r = r_0$ of the cylinder, but they are adequate for most requirements.

Equations (4.22) demonstrate that E_{el} depends on the core radius r_0 and the crystal radius R, but only logarithmically. E_{el} (edge) is greater than E_{el} (screw) by $1/(1-\nu) \approx 3/2$. Taking $R = 1$ mm, $r_0 = 1$ nm, $G = 40$ GN m^{-2} and $b = 0.25$ nm, the elastic strain energy of an edge dislocation will be about 4 nJ m^{-1} or about 1 aJ(6 eV) for each atom plane threaded by the dislocation. In crystals containing many dislocations, the dislocations tend to form in configurations in which the superimposed long-range elastic fields cancel. The energy per dislocation is thereby reduced and an appropriate value of R is approximately half the average spacing of the dislocations arranged at random.

Estimates of the energy of the core of the dislocation are necessarily very approximate. However, the estimates that have been made suggest that the core energy will be of the order of 1 eV for each atom plane threaded by the dislocation, and is thus only a small fraction of the elastic energy. However, in contrast to the elastic energy, the energy of the core will vary as the dislocation moves through the crystal and this gives rise to the lattice resistance to dislocation motion discussed in section 10.3.

The validity of elasticity theory for treating dislocation energy outside a core region has been demonstrated by atomic scale computer simulation (section 2.4). Figure 4.8 shows data for an atomic model of alpha iron containing a straight edge dislocation with Burgers vector $\frac{1}{2}[111]$ and line direction $[11\bar{2}]$. E_{total} is the strain energy within a cylinder of radius R with the dislocation along its axis. The energy varies logarithmically with R, as predicted by equation (4.22), outside a core of radius 0.7 nm, which is about $2.6b$. The core energy is about 7 eV nm^{-1}.

It was mentioned in the preceding section that the elastic field of a *mixed* dislocation (see Fig. 3.8(b)) is the superposition of the fields of its edge and screw parts. As there is no interaction between them, the total elastic energy is simply the sum of the edge and screw energies with b replaced by $b \sin \theta$ and $b \cos \theta$ respectively:

$$E_{el}(\text{mixed}) = \left[\frac{Gb^2 \sin^2\theta}{4\pi(1-\nu)} + \frac{Gb^2 \cos^2\theta}{4\pi} \right] \ln\left(\frac{R}{r_0}\right) = \frac{Gb^2(1-\nu\cos^2\theta)}{4\pi(1-\nu)} \ln\left(\frac{R}{r_0}\right)$$

(4.23)

which falls between the energy of an edge and a screw dislocation.

From the expressions for edge, screw and mixed dislocations it is clear that the energy per unit length is relatively insensitive to the character of the dislocation and also to the values of R and r_0. Taking realistic values for R and r_0 all the equations can be written approximately as

$$E_{el} = \alpha Gb^2 \qquad (4.24)$$

FIGURE 4.9
Reaction of two dislocations to form a third.

where $\alpha \approx 0.5-1.0$. This leads to a very simple rule (*Frank's rule*) for determining whether or not it is energetically feasible for two dislocations to react and combine to form another. Consider the two dislocations in Fig. 4.9 with Burgers vectors $\mathbf{b}_1$ and $\mathbf{b}_2$ given by the Burgers circuit construction (section 1.4). Allow them to combine to form a new dislocation with Burgers vector $\mathbf{b}_3$ as indicated. From equation (4.24), the elastic energy per unit length of the dislocations is proportional to b_1^2, b_2^2 and b_3^2 respectively. Thus, if $(b_1^2 + b_2^2) > b_3^2$, the reaction is favorable for it results in a reduction in energy. If $(b_1^2 + b_2^2) < b_3^2$, the reaction is unfavorable and the dislocation with Burgers vector b_3 is liable to dissociate into the other two. If $(b_1^2 + b_2^2) = b_3^2$, there is no energy change. These three conditions correspond to the angle ϕ in Fig. 4.9 satisfying $\pi/2 < \phi \le \pi$, $0 \le \phi < \pi/2$ and $\phi = \pi/2$ respectively. In this argument the assumption is made that there is no additional interaction energy involved, i.e. that before and after the reaction the reacting dislocations are separated sufficiently so that the interaction energy is small. If this is not so, the reactions are still favorable and unfavorable, but the energy changes are smaller than implied above. Frank's rule is used to consider the feasibility of various dislocation reactions in Chapters 5 and 6.

4.5 FORCES ON DISLOCATIONS

When a sufficiently high stress is applied to a crystal containing dislocations, the dislocations move and produce plastic deformation either by slip as described in section 3.3 or, at sufficiently high temperatures, by climb (section 3.6). The load producing the applied stress therefore does work on the crystal when a dislocation moves, and so the dislocation responds to the stress as though it experiences a force equal to the work done divided by the distance it moves. The force defined in this way is a virtual, rather than real, force, but the force concept is useful for treating the mechanics of dislocation behavior. The *glide* force is considered in this section and the climb force in section 4.7.

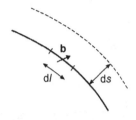

FIGURE 4.10

The displacement ds used to determine the glide force on an element dl in its glide plane.

Consider a dislocation moving in a slip plane under the influence of a uniform resolved shear stress τ (Fig. 4.10). When an element dl of the dislocation line of Burgers vector $\mathbf{b}$ moves forward a distance ds the crystal planes above and below the slip plane will be displaced relative to each other by b. The average shear displacement of the crystal surface produced by glide of dl is

$$\left(\frac{ds\,dl}{A}\right)b \qquad (4.25)$$

where A is the area of the slip plane. The external force due to τ acting over this area is $A\tau$, so that the work done when the element of slip occurs is

$$dW = A\tau\left(\frac{ds\,dl}{A}\right)b \qquad (4.26)$$

The glide force F on a unit length of dislocation is defined as the work done when unit length of dislocation moves unit distance. Therefore

$$F = \frac{dW}{ds\,dl} = \frac{dW}{dA} = \tau b \qquad (4.27)$$

The stress τ *is the shear stress in the glide plane resolved in the direction of* $\mathbf{b}$ and the glide force F acts normal to the dislocation at every point along its length, irrespective of the line direction. The positive sense of the force is given by the physical reasoning of section 3.3.

In addition to the force due to an externally applied stress, a dislocation has a *line tension* which is analogous to the surface tension of a soap bubble or a liquid. This arises because, as outlined in the previous section, the strain energy of a dislocation is proportional to its length and an increase in length results in an increase in energy. The line tension has units of energy per unit length. From the approximation used in equation (4.24), the line tension, which may be defined as *the increase in energy per unit increase in the length of a dislocation line*, is

$$T = \alpha Gb^2 \qquad (4.28)$$

Consider the curved dislocation in Fig. 4.11. The line tension will produce forces tending to straighten the line and so reduce the total energy. The direction of the net force is perpendicular to the dislocation and towards the center of curvature. The line will only remain curved if there is a shear stress which produces a force on the dislocation line in the opposite sense. The shear stress τ_0 needed to maintain a radius of curvature R is found in the following way. The angle subtended at the center of curvature is $d\theta = dl/R$, assumed to be $\ll 1$. The outward force along OA due to the applied stress acting on the elementary piece of dislocation is $\tau_0 b\,dl$ from equation (4.27), and the opposing inward force along OA due to the line tension T at the ends of the element is $2T \sin(d\theta/2)$, which is equal to $T\,d\theta$ for small values of $d\theta$. The line will be in equilibrium in this curved position when

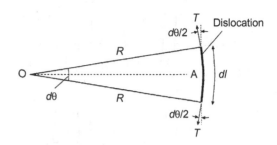

FIGURE 4.11
Curved element of dislocation under line tension forces T.

$$T\,d\theta = \tau_0 b\,dl$$

$$\text{i.e. } \tau_0 = \frac{T}{bR} \qquad (4.29)$$

Substituting for T from equation (4.28)

$$\tau_0 = \frac{\alpha G b}{R} \qquad (4.30)$$

This gives an expression for the stress required to bend a dislocation to a radius R and is used many times in subsequent chapters. A particularly direct application is in the understanding of the Frank–Read dislocation multiplication source described in Chapter 8.

Equation (4.30) assumes from equation (4.24) that edge, screw and mixed segments have the same energy per unit length, and the curved dislocation of Fig. 4.11 is therefore the arc of a circle. This is only strictly valid if Poisson's ratio ν equals zero. In all other cases, the line experiences a torque tending to rotate it towards the screw orientation where its energy per unit length is lower. The true line tension of a mixed segment is

$$T = E_{el}(\theta) + \frac{d^2 E_{el}(\theta)}{d\theta^2} \qquad (4.31)$$

where $E_{el}(\theta)$ is given by equation (4.23). T for a screw segment is four times that of an edge when $\nu = 1/3$. Thus, for a line bowing under a uniform stress, the radius of curvature at any point is still given by equation (4.29), but the overall line shape is approximately elliptical with major axis parallel to the Burgers vector; the axial ratio is approximately $1/(1-\nu)$. For most calculations, however, equation (4.30) is an adequate approximation.

4.6 FORCES BETWEEN DISLOCATIONS

A simple semi-qualitative argument will illustrate the significance of the concept of a force between dislocations. Consider two parallel edge dislocations lying in the same slip plane. They can either have the same sign as in

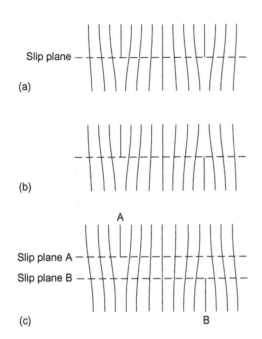

FIGURE 4.12

Arrangement of edge dislocations with parallel Burgers vectors lying in parallel slip planes. (a) Like dislocations on the same slip plane, (b) unlike dislocations on the same slip plane, and (c) unlike dislocations on slip planes separated by a few atomic spacings.

Fig. 4.12 (a) or opposite sign as in Fig. 4.12 (b). When the dislocations are separated by a large distance the total elastic energy per unit length of the dislocations in both situations will be, from equation (4.24)

$$\alpha Gb^2 + \alpha Gb^2 \tag{4.32}$$

When the dislocations in Fig. 4.12(a) are very close together the arrangement can be considered approximately as a single dislocation with a Burgers vector magnitude $2b$ and the elastic energy will be given by

$$\alpha G(2b)^2 \tag{4.33}$$

which is twice the energy of the dislocations when they are separated by a large distance. Thus the dislocations will tend to repel each other to reduce their total elastic energy. When dislocations of opposite sign (Fig. 4.12(b)) are close together, the effective magnitude of their Burgers vectors will be zero, and the corresponding long-range elastic energy zero also. Thus dislocations of opposite sign will attract each other to reduce their total elastic energy. The positive and negative edge dislocations in Fig. 4.12(b) will combine and annihilate each other. These conclusions regarding repulsion and attraction also follow for dislocations of mixed orientation from Frank's rule by putting $\phi = 0$ or π in Fig. 4.9. Similar effects occur when the two dislocations do not lie in the same slip plane (Fig. 4.12(c)), but the conditions for attraction and repulsion are usually more complicated, as discussed below.

The basis of the method used to obtain the force between two dislocations is the determination of the additional work done in introducing the second dislocation into a crystal which already contains the first. Consider two dislocations lying parallel to the z-axis in Fig. 4.13(a): they are shown as edge dislocations for simplicity. The total energy of the system consists of the self-energy of dislocation I plus the self-energy of dislocation II plus the elastic interaction energy between I and II. The *interaction energy* E_{int} is the work done in displacing the faces of the cut which creates II in the presence of the stress field of I. The displacements across the cut are b_x, b_y, b_z, the components of the Burgers vector **b** of II. By visualizing the cut parallel to either the x or y axes, two alternative expressions for E_{int} per unit length of II are

$$E_{int} = +\int_x^\infty (b_x\sigma_{xy} + b_y\sigma_{yy} + b_z\sigma_{zy})dx$$
$$E_{int} = -\int_y^\infty (b_x\sigma_{xx} + b_y\sigma_{yx} + b_z\sigma_{zx})dy \tag{4.34}$$

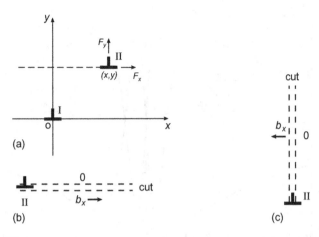

FIGURE 4.13

(a) Forces on dislocation II due to interaction with dislocation I. Dislocations shown as edges for simplicity. (b),(c) x component of displacement of cut faces used to create dislocation II.

where the stress components are those due to I on the cut face of II. (The signs of the right-hand side of these equations arise because if the displacements of **b** are taken to occur on the face of a cut with outward normal in the positive y and x directions, respectively, they are in the direction of positive x, y, z for the first case (x-axis cut) and negative x, y, z for the second (y-axis cut) as shown explicitly in Figs 4.13(b) and (c), respectively, for the b_x component.)

The interaction force on II is obtained simply by differentiation of these expressions, i.e. $F_x = -\partial E_{int}/\partial x$ and $F_y = -\partial E_{int}/\partial y$. For the two *parallel edge dislocations* with parallel Burgers vectors shown in Fig. 4.13(a), $b_y = b_z = 0$ and $b_x = b$, and the components of the force per unit length acting on II are therefore

$$F_x = \sigma_{xy}b \quad F_y = -\sigma_{xx}b \tag{4.35}$$

where σ_{xy} and σ_{xx} are the stresses of I evaluated at position (x, y) of II. The forces are reversed if II is a negative edge i.e. the dislocations have opposite sign. Equal and opposite forces act on I. F_x is the force in the glide direction and F_y the force perpendicular to the glide plane. Substituting from equation (4.16) gives

$$F_x = \frac{Gb^2}{2\pi(1-\nu)} \frac{x(x^2 - y^2)}{(x^2 + y^2)^2}$$

$$F_y = \frac{Gb^2}{2\pi(1-\nu)} \frac{y(3x^2 - y^2)}{(x^2 + y^2)^2} \tag{4.36}$$

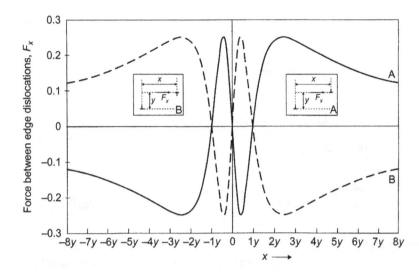

FIGURE 4.14

Glide force per unit length between parallel edge dislocations with parallel Burgers vectors from equation (4.36). Unit of force F_x is $Gb^2/2\pi(1-\nu)y$. The full curve A is for like dislocations and the broken curve B for unlike dislocations.

Since an edge dislocation can move by slip only in the plane contained by the dislocation line and its Burgers vector, the component of force which is most important in determining the behavior of the dislocations in Fig. 4.13(a) is F_x. For dislocations of the same sign, inspection of the variation of F_x with x reveals the following:

F_x	nature	x range
negative	repulsive	$-\infty < x < -y$
positive	attractive	$-y < x < 0$
negative	attractive	$0 < x < y$
positive	repulsive	$y < x < \infty$

The sign and nature of F_x is reversed if I and II are edge dislocations of opposite sign. F_x is plotted against x, expressed in units of y, in Fig. 4.14. It is zero when $x = 0$, $\pm y$, $\pm\infty$, but of these, the positions of stable equilibrium are seen to be $x = 0$, $\pm\infty$ for edges of the same sign and $\pm y$ if they have the opposite sign.

It follows that an array of edge dislocations of the same sign is most stable when the dislocations lie vertically above one another as in Fig. 4.15(a).

FIGURE 4.15

Stable positions for two edge dislocations of (a) the same sign and (b) opposite sign.

This is the arrangement of dislocations in a small angle pure tilt boundary described in Chapter 9. Furthermore, edge dislocations of opposite sign gliding past each other on parallel slip planes tend to form stable *dipole* pairs as in Fig. 4.15(b) at low applied stresses (section 10.8).

Comparison of the glide force F_x in equation (4.35) with F in equation (4.27) shows that since σ_{xy} is the shear stress in the glide plane of dislocation II acting in the direction of its Burgers vector, *equation (4.27) holds for both external and internal sources of stress.*

Consider two parallel screw dislocations (Fig. 4.16), one lying along the z-axis. The radial and tangential components of force on the other are

$$F_r = \sigma_{z\theta} b \quad F_\theta = \sigma_{zr} b \tag{4.37}$$

and substituting from equations (4.15)

$$F_r = Gb^2/2\pi r \quad F_\theta = 0 \tag{4.38}$$

The force is much simpler in form than that between two edge dislocations because of the radial symmetry of the screw field. F_r is repulsive for screws of the same sign and attractive for screws of opposite sign. It is readily shown from either equations (4.35) or equations (4.37) that no forces act between a pair of parallel dislocations consisting of a pure edge and a pure screw, as expected from the lack of mixing of their stress fields (see section 4.3).

FIGURE 4.16

Forces considered for the interaction between two screw dislocations.

4.7 CLIMB FORCES

The force component F_y in equation (4.35) is a *climb force* per unit length resulting from the normal stress σ_{xx} of dislocation I attempting to squeeze the extra half-plane of II from the crystal. This can only occur physically if intrinsic point defects can be emitted or absorbed at the dislocation core

Mechanical force F

$c < c_0$ $c > c_0$

Chemical force f

$c < c_0$ $c > c_0$

FIGURE 4.17

Mechanical and chemical forces for climb of an edge dislocation. Vacancies (shown as □) have a local concentration c in comparison with the equilibrium concentration c_0 in a dislocation-free crystal.

of II (see section 3.6). As in the case of glide forces, climb forces can arise from external and internal sources of stress. The former are important in *creep*, and the latter provided the example of conservative climb in section 3.8 (Fig. 3.24). Line tension can also produce climb forces, but in this case the force acts to reduce the line length in the extra half-plane: shrinkage of prismatic loops as in Fig. 3.20 is an example. However, since the creation and annihilation of point defects are involved in climb, *chemical forces* due to defect concentration changes must be taken into account in addition to these *mechanical forces*.

It was seen in section 3.6 that when an element l of dislocation is displaced through s, the local volume change is $\mathbf{b} \times \mathbf{l} \cdot \mathbf{s}$. Consider a segment length l of a positive edge dislocation climbing upwards through distance s in response to a mechanical climb force F per unit length. The work done is Fls and the number of vacancies absorbed is bls/Ω, where Ω is the volume per atom. The vacancy formation energy is therefore changed by $F\Omega/b$. As a result of this chemical potential, the equilibrium vacancy concentration at temperature T in the presence of the dislocation is reduced to

$$c = \exp[-(E_f^v + F\Omega/b)/kT]$$
$$= c_0 \exp[-F\Omega/bkT] \tag{4.39}$$

where c_0 is the equilibrium concentration in a stress-free crystal (equation (1.3)). For negative climb involving vacancy emission ($F < 0$) the sign of the chemical potential is changed so that $c > c_0$. Thus, the vacancy concentration deviates from c_0, building up a chemical force per unit length on the line

$$f = \frac{bkT}{\Omega} \ln(c/c_0) \tag{4.40}$$

until f balances F in equilibrium. Conversely, in the presence of a supersaturation c/c_0 of vacancies, the dislocation climbs up under the chemical force f until compensated by, say, external stresses or line tension. The latter is used in the analysis for a dislocation climb source in section 8.7. The nature of these forces is illustrated schematically in Fig. 4.17. By substituting reasonable values of T and Ω in equation (4.40), it is easy to show that even moderate supersaturations of vacancies can produce forces much greater than those arising from external stresses.

The rate of climb of a dislocation in practice depends on (a) the direction and magnitude of the mechanical and chemical forces, F and f, (b) the

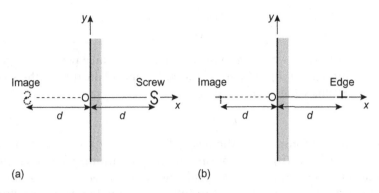

FIGURE 4.18
(a) Screw and (b) edge dislocations a distance d from a surface $x = 0$. The image dislocations are in space a distance d from the surface.

mobility of jogs (section 3.6) and (c) the rate of migration of vacancies through the lattice to or from the dislocation.

4.8 IMAGE FORCES

A dislocation near a surface experiences forces not encountered in the bulk of a crystal. The dislocation is attracted towards a free surface because the material is effectively more compliant there and the dislocation energy is lower: conversely, it is repelled by a rigid surface layer. To treat this mathematically, extra terms must be added to the infinite-body stress components given in section 4.3 in order that the required surface conditions are satisfied. When evaluated at the dislocation line, as in equations (4.35) and (4.37), they result in a force. The analysis for infinite, straight dislocation lines parallel to the surface is relatively straightforward.

Consider screw and edge dislocations parallel to, and distance d from, a surface $x = 0$ (Fig. 4.18); the edge dislocation has Burgers vector $\mathbf{b}$ in the x direction. For a *free* surface, the tractions σ_{xx}, σ_{yx} and σ_{zx} must be zero on the plane $x = 0$. Consideration of equation (4.13) shows that these boundary conditions are met for the screw if the infinite-body result is modified by adding to it the stress field of an imaginary screw dislocation of opposite sign at $x = -d$ (Fig. 4.18(a)). The required solution for the stress in the body ($x > 0$) is therefore

$$\sigma_{zx} = \frac{-Ay}{(x_-^2 + y^2)} + \frac{Ay}{(x_+^2 + y^2)} \tag{4.41}$$

$$\sigma_{zy} = \frac{Ax_-}{(x_-^2 + y^2)} - \frac{Ax_+}{(x_+^2 + y^2)} \qquad (4.42)$$

where $x_- = (x-d)$, $x_+ = (x+d)$ and $A = Gb/2\pi$. The force per unit length in the x-direction $F_x(= \sigma_{zy}b)$ induced by the surface is obtained from the second term in σ_{zy} evaluated at $x = d$, $y = 0$. It is

$$F_x = -Gb^2/4\pi d \qquad (4.43)$$

and is simply the force due to the *image* dislocation at $x = -d$. For the edge dislocation (Fig. 4.18(b)), superposing the field of an imaginary edge dislocation of opposite sign at $x = -d$ annuls the stress σ_{xx} on $x = 0$, but not σ_{yx}. When the extra terms are included to fully match the boundary conditions, the shear stress in the body is found to be

$$\sigma_{yx} = \frac{Dx_-(x^2 - y^2)}{(x_-^2 + y^2)^2} - \frac{Dx_+(x_+^2 - y^2)}{(x_+^2 + y^2)^2} - \frac{2Dd[x_- x_+^3 - 6xx_+ y^2 + y^4]}{(x_+^2 + y^2)^3} \qquad (4.44)$$

where $D = A/(1-\nu)$. The first term is the stress in the absence of the surface, the second is the stress appropriate to an image dislocation at $x = -d$, and the third is that required to make $\sigma_{yx} = 0$ when $x = 0$. The force per unit length $F_x(= \sigma_{yx}b)$ arising from the surface is given by putting $x = d$, $y = 0$ in the second and third terms. The latter contributes zero, so that the force is

$$F_x = -Gb^2/4\pi(1-\nu)d \qquad (4.45)$$

and is again equivalent to the force due to the image dislocation.

The *image forces* decrease slowly with increasing d and are capable of removing dislocations from near-surface regions. They are important, for example, in specimens for transmission electron microscopy (section 2.2) when the slip planes are orientated at large angles ($\simeq 90°$) to the surface. It should be noted that a second dislocation near the surface would experience a force due to its own image *and* the surface terms in the field of the first. The interaction of dipoles, loops and curved dislocations with surfaces is therefore complicated, and only given approximately by images.

FURTHER READING

Bacon DJ, Barnett DM, Scattergood RO: Anisotropic continuum theory of lattice defects, *Prog Mater Sci* 23:51, 1979.

Cottrell AH: *Dislocations and plastic flow in crystals*, 1953, Oxford University Press.

Friedel J: *Dislocations*, 1964, Pergamon.

Hirth JP, Lothe J: *Theory of dislocations*, 1982, Wiley.

Lardner RW: *Mathematical theory of dislocations and fracture*, 1974, Univ. of Toronto Press.

Mura T: *Micromechanics of defects in solids*, 1982, Martinus Nijhoff.

Nabarro FRN: *The theory of crystal dislocations*, 1967, Oxford University Press.

Nye JF: *Physical properties of crystals*, 1967, Oxford University Press.

Seeger A: *Theory of lattice imperfections, Handbuch der physik*, Vol VII, 1955, Springer-Verlag, part 1, p. 383.

Steeds JW: *Anisotropic elasticity theory of dislocations*, 1973, Oxford University Press.

Teodosiu C: *Elastic models of crystal defects*, 1982, Springer.

Timoshenko SP, Goodier JN: *Theory of elasticity*, 1970, McGraw-Hill.

Weertman J, Weertman JR: *Elementary dislocation theory*, 1993, Oxford University Press, (1st edition 1964, Macmillan).

CHAPTER 5

Dislocations in Face-centered Cubic Metals

5.1 PERFECT DISLOCATIONS

Many common metals such as copper, silver, gold, aluminum, nickel and their alloys, have a face-centered cubic crystal structure (Fig. 1.7). The pure metals are soft, with critical resolved shear stress values for single crystals $0.1 - 1 \, \mathrm{MN \, m^{-2}}$. They are ductile but can be hardened considerably by plastic deformation and alloying. The deformation behavior is closely related to the atomic structure of the core of dislocations, which is more complex than that described in Chapters 1 and 3.

The shortest lattice vectors, and therefore the most likely Burgers vectors for dislocations in the face-centered cubic structure, are of the type $\frac{1}{2}\langle 110 \rangle$ and $\langle 001 \rangle$. Since the energy of a dislocation is proportional to the square of the magnitude of its Burgers vector b^2 (section 4.4), the energy of $\frac{1}{2}\langle 110 \rangle$ dislocations in an isotropic solid will be only half that of $\langle 001 \rangle$, i.e. $2a^2/4$ compared with a^2. Thus, $\langle 001 \rangle$ dislocations are much less favored energetically and, in fact, are only rarely observed. Since $\frac{1}{2}\langle 110 \rangle$ is a translation vector for the lattice, glide of a dislocation with this Burgers vector leaves behind a perfect crystal and the dislocation is a *perfect dislocation*. Figure 5.1 represents a $\frac{1}{2}[110]$ edge dislocation in a face-centered cubic metal. The (110) planes perpendicular to **b** are illustrated and have a two-fold stacking sequence *ABAB...* (section 1.2). The 'extra half-plane' consists of two (110) half-planes in the *ABAB...* sequence. Movement of this unit dislocation by glide retains continuity of the *A* planes and the *B* planes across the glide plane, except at the dislocation core where the extra half-planes terminate.

5.2 PARTIAL DISLOCATIONS – THE SHOCKLEY PARTIAL

Examination of Fig. 5.1 suggests that the two extra (110) planes need not necessarily be immediately adjacent to each other. If separated, each would

Introduction to Dislocations.

FIGURE 5.1

Unit edge dislocation $\frac{1}{2}[110]$ in a face-centered cubic crystal. *(After Seeger (1957), Dislocations and Mechanical Properties of Crystals, p. 243, Wiley.)*

appear as a single dislocation with a Burgers vector shorter than $\frac{1}{2}\langle 110 \rangle$. The perfect dislocation would split into two *partial dislocations*, as explained in section 5.3.

From the description of dislocation movement in Chapter 3, it may be deduced that motion of a dislocation whose Burgers vector is not a lattice vector leaves behind an imperfect crystal containing a stacking fault. Thus, when a stacking fault ends inside a crystal, the boundary in the plane of the fault, separating the faulted region from the perfect region of the crystal, is a partial dislocation. Two of the important partial dislocations, recognized in face-centered cubic metals, are the *Shockley partial*, which is associated with slip, and the *Frank partial* (see section 5.5). The formation of a Shockley partial edge dislocation is illustrated in Fig. 5.2 and can be compared with the formation of an edge dislocation in an elastic model (Fig. 3.3). The diagram represents a $(10\bar{1})$ section through the lattice. The close-packed (111) planes lie at right angles to the plane of the diagram. At the right of the diagram the (111) layers are stacked in the sequence $ABCABC\ldots$ and the lattice is perfect. At the left of the diagram the A layer atoms above LM have slipped in the $[1\bar{2}1]$

FIGURE 5.2

Formation of a $\frac{1}{6}[1\bar{2}1]$ Shockley partial dislocation at M due to slip along LM. The open circles represent the positions of atoms in the $(10\bar{1})$ plane of the diagram and the filled circles the positions of the atoms in the $(10\bar{1})$ planes immediately above and below the plane of the diagram. Some lattice vectors that lie in the $(10\bar{1})$ plane are indicated. *(After Read (1953), Dislocations in Crystals, McGraw-Hill.)*

direction to a B layer position, the B atoms have slipped to C and the C atoms have slipped to A. This has produced a stacking fault and a partial dislocation. The fault vector, which is in the (111) slip plane, is $\mathbf{b} = \frac{1}{6}[1\bar{2}1]$, and the magnitude of the vector is $a/\sqrt{6}$; this compares with $a/\sqrt{2}$ for the $\frac{1}{2}\langle 110 \rangle$ perfect dislocation, as shown by the following:

$$\text{Shockley}\left(\mathbf{b} = \frac{1}{6}\langle 112 \rangle\right) : b^2 = \frac{a^2}{36}(1^2 + 1^2 + 2^2) = \frac{a^2}{6}$$

$$(5.1)$$

$$\text{Perfect}\left(\mathbf{b} = \frac{1}{2}\langle 110 \rangle\right) : b^2 = \frac{a^2}{4}(1^2 + 1^2 + 0) = \frac{a^2}{2}$$

The Burgers vector of a partial dislocation is described in the same way as that of a perfect dislocation, except that the Burgers circuit must start and finish in the surface of the stacking fault; if the circuit started at any other position it would be necessary to cross the fault plane and the one-to-one correspondence of the circuits in the perfect and imperfect lattices would not be maintained. Since the Burgers vector of a partial dislocation is not a unit lattice vector, the 'finish' position of the circuit in the perfect lattice (Fig. 1.19(b)) is not a lattice site.

5.3 SLIP

Slip occurs between close-packed {111} atomic planes and the observed slip direction is $\langle 110 \rangle$. Only rarely does glide occur on other planes. Since slip involves the sliding of close-packed planes of atoms over each other, a simple experiment can be made to see how this can occur. The close-packed planes can be simulated by a set of hard spheres, as illustrated schematically in Fig. 5.3, and have a three-fold stacking sequence ABCABC... (see also Fig. 1.7).

One layer is represented by the full circles, A, the second identical layer rests in the sites marked B and the third takes the positions C. Consider the movement of the layers when they are sheared over each other to produce a displacement in the slip direction. It will be found that the B layer of atoms, instead of moving from one B site to the next B site over the top of the A atoms (vector $\mathbf{b}_1$), will move first to the nearby C site along the 'valley' between the two A

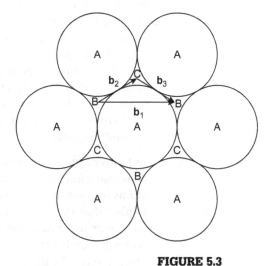

FIGURE 5.3

Slip of {111} planes in face-centered cubic metals.

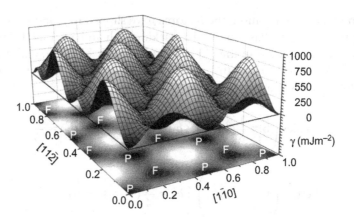

FIGURE 5.4

γ surface for the (111) plane of copper obtained by atomic-scale computer simulation. γ is zero for perfect stacking (P). Intrinsic fault (F) is metastable (with $\gamma \simeq 40$ mJ m^{-2}) and corresponds to fault vectors of the form $\frac{1}{6}\langle 112 \rangle$. *(Courtesy Yu. N. Osetsky.)*

atoms (vector $\mathbf{b}_2$) and then to the new B site via a second valley (vector $\mathbf{b}_3$). Thus, the B plane will slide over the A plane in a zig-zag motion.

This simple hard-sphere description has been confirmed by computer simulation (section 2.4). As one part of a crystal slides over another across a {111} plane to form a stacking fault, the energy varies from minima at translations corresponding to perfect stacking to maxima when atoms across the fault are directly over each other. The extra crystal energy per unit area of stacking fault is the *stacking fault energy* γ. A plot of γ versus fault translation vector is a γ (or *gamma*) *surface* and is shown for a model of copper in Fig. 5.4. The positions of perfect stacking ($\gamma = 0$) are labeled P and are connected by translations of the type $\frac{1}{2}\langle 110 \rangle$. The locations F correspond to the *intrinsic stacking fault* (section 1.3) and are reached from P by vectors of the form $\frac{1}{6}\langle 112 \rangle$. It can be seen that these translations follow low energy paths. Furthermore, F is a local minimum (with $\gamma \simeq 40$ mJ m^{-2} for copper) so that the fault is stable.

In terms of glide of a perfect dislocation with Burgers vector $\mathbf{b}_1 = \frac{1}{2}\langle 110 \rangle$, this demonstration suggests that it will be energetically more favorable for the B atoms to move to B via the C positions, i.e. P to F to P on the γ surface. This implies that the dislocation glides as two partial dislocations, one immediately after the other. The first has Burgers vector $\mathbf{b}_2$ and the second Burgers vector $\mathbf{b}_3$, each of which has the form $\frac{1}{6}\langle 112 \rangle$. The perfect dislocation with Burgers vector $\mathbf{b}_1$ therefore *splits up* or *dissociates* into two dislocations $\mathbf{b}_2$ and $\mathbf{b}_3$ according to the reaction:

$$\mathbf{b}_1 \rightarrow \mathbf{b}_2 + \mathbf{b}_3 \qquad (5.2)$$

or

$$\frac{1}{2}\langle 110\rangle \rightarrow \frac{1}{6}\langle 211\rangle + \frac{1}{6}\langle 12\bar{1}\rangle$$

In any of the four {111} slip planes there are three $\langle 110\rangle$ directions and three $\langle 112\rangle$ directions, as shown for the (111) plane in Fig. 5.5. Dissociation of a perfect dislocation into two partials on, say, the (111) plane can therefore be one of the Burgers vector reactions given by

$$\frac{1}{2}[\bar{1}10] \rightarrow \frac{1}{6}[\bar{2}11] + \frac{1}{6}[\bar{1}2\bar{1}]$$

$$\frac{1}{2}[\bar{1}01] \rightarrow \frac{1}{6}[\bar{2}11] + \frac{1}{6}[\bar{1}\bar{1}2] \qquad (5.3)$$

$$\frac{1}{2}[0\bar{1}1] \rightarrow \frac{1}{6}[1\bar{2}1] + \frac{1}{6}[\bar{1}\bar{1}2]$$

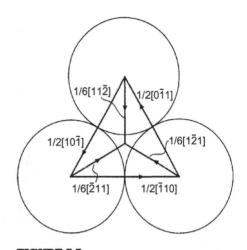

FIGURE 5.5
Burgers vectors of perfect and Shockley partial dislocations in the (111) plane.

and the reactions produced by complete reversal of each vector. It is necessary in dislocation reactions to ensure that the total Burgers vector is unchanged, as explained in section 1.4. The right-hand side of the first of reactions (5.3), for example, is $\frac{1}{6}[\bar{2}+\bar{1},1+2,1+\bar{1}]$, which equals the left-hand side. The same result is demonstrated diagrammatically by the vector triangles in Fig. 5.5.

Frank's rule (section 4.4) shows that the splitting reaction (5.2) is energetically favorable, for from (5.1) $b_1^2 = a^2/2$ which is greater than $b_2^2 + b_3^2 = a^2/3$. Since the two Burgers vectors $\mathbf{b}_2$ and $\mathbf{b}_3$ are at 60° to each other, the two partial dislocations repel each other with a force due to their elastic interaction (see section 4.6). The force may be calculated from the separate forces between their screw components (equation (4.38)) and their edge components (equation (4.36) with $y = 0$). If the spacing of the partials is d, the repulsive force per unit length between the partials of either pure edge or pure screw perfect dislocations is

$$F = \frac{Gb^2(2 + \nu)}{8\pi(1 - \nu)d} \quad \text{(edge)}$$

$$F = \frac{Gb^2(2 - 3\nu)}{8\pi(1 - \nu)d} \quad \text{(screw)}$$

$$(5.4)$$

respectively, where $b \, (= a/\sqrt{6})$ is the magnitude of $\mathbf{b}_2$ and $\mathbf{b}_3$. For the special case $\nu = 0$, F is the same in all line orientations:

$$F = \frac{G\mathbf{b}_2 \cdot \mathbf{b}_3}{2\pi d} = \frac{Gb^2}{4\pi d} \qquad (5.5)$$

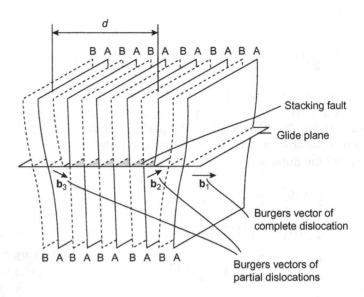

FIGURE 5.6

Formation of an extended dislocation by dissociation of a unit edge dislocation (Fig. 5.1) into two Shockley partials of Burgers vectors $\mathbf{b}_2$ and $\mathbf{b}_3$ separated by a stacking fault. (The labels ABAB... refer to the stacking of (110) planes, as in fig. 5.2.) *(After Seeger (1957), Dislocations and Mechanical Properties of Crystals p. 243 Wiley.)*

and this is a reasonable approximation to the more general results. Since $\mathbf{b}_2$ and $\mathbf{b}_3$ are the Burgers vectors of Shockley partial dislocations, it follows that if they separate there will be a *ribbon* of stacking fault between them. The stacking sequence of {111} planes will be perfect outside the dislocation (ABCABCABC...) and faulted between the partial dislocations (ABCACABC...). This is the intrinsic fault discussed above and is equivalent to four layers of close-packed hexagonal stacking in a face-centered cubic crystal. The stacking fault energy $(J\,m^{-2}$ or $N\,m^{-1})$ provides a force γ per unit length of line $(N\,m^{-1})$ tending to pull the dislocations together. An equilibrium separation will be established when the repulsive and attractive forces balance. The approximate equilibrium separation is obtained by equating γ to F in equation (5.5).

$$d = \frac{Gb^2}{4\pi\gamma} \tag{5.6}$$

The unit edge dislocation illustrated in Fig. 5.1 will split up as illustrated in Fig. 5.6. The configuration is called an *extended dislocation*. Note that the width, d, is inversely proportional to the stacking fault energy.

Screw dislocations can form a similar configuration to Fig. 5.6. Except for the size of the partial separation d, the dissociation of a perfect dislocation is independent of its character (edge, screw or mixed). During glide under stress, a dissociated dislocation moves as a pair of partials bounding the fault ribbon, the leading partial creating the fault and the trailing one removing it: the total slip vector is $\mathbf{b}_1 = \frac{1}{2}\langle 110 \rangle$. Experimental observations of extended dislocations in thin foils have confirmed that this geometry is correct. Figure 5.7 shows sets of extended dislocations lying in parallel slip planes. The stacking fault ribbon between two partials appears as a parallel fringe pattern. The individual partials are not always visible and their positions are illustrated in Fig. 5.7(b). Numerous estimates of γ have been made from direct observation of the spacing of partial dislocations in the transmission electron microscope (see sections 2.2 and 7.8), from the shrinkage rate of faulted prismatic loops such as those shown in Fig. 3.20, and indirectly from the temperature dependence of the flow stress of single crystals. It is probable that $\gamma \approx 140 \text{ mJ m}^{-2}$ for aluminum, $\approx 40 \text{ mJ m}^{-2}$ for copper and $\approx 20 \text{ mJ m}^{-2}$ for silver. The corresponding width of the stacking-fault ribbon given by equation (5.6) is about a, $5a$ and $7a$, respectively. The widths predicted by equations (5.4) are rather greater than these for edge dislocations and less for screws. For comparison with the sketch in Fig. 5.6, the atomic positions around the dissociated dislocation in a model crystal of copper as obtained by computer simulation (see section 2.4) are shown in Fig. 5.8.

Cross slip (see section 3.4) is more difficult to achieve when dissociation occurs, for a $\frac{1}{6}\langle 112 \rangle$ vector lies in only one $\{111\}$ plane and so an individual Shockley partial cannot cross slip. An extended dislocation is therefore constrained to glide in the $\{111\}$ plane of its fault. Although extended screw dislocations cannot cross slip it is possible to form a *constriction* in the screw dislocation and then the perfect dislocation at the constriction can glide in another $\{111\}$ plane as in Fig. 3.9. A constriction is illustrated in Fig. 5.9. Energy is required to form a constriction, since the dislocation is in its lowest energy state when dissociated, and this occurs more readily in metals with a high stacking fault energy such as aluminum. It follows that cross slip will be most difficult in metals with a low stacking fault energy and this produces significant effects on the deformation behavior. Formation of a constriction

(a)

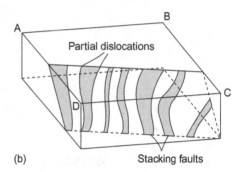

(b)

FIGURE 5.7

(a) Transmission electron micrograph of extended dislocations in a copper−7 per cent aluminum alloy. (*From Howie*, Metallurgical Reviews, *6, 467, 1961.*) (b) Arrangement of dislocations in the inset in (a).

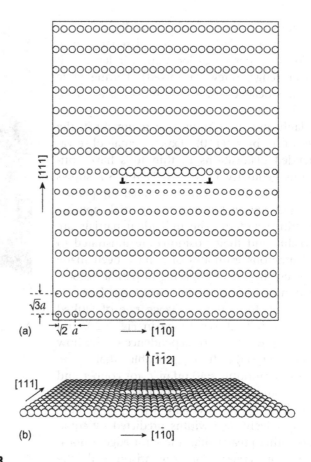

(a)

$\sqrt{2}\,a$ $[1\bar{1}0]$

$[1\bar{1}2]$

[111]

(b) $[1\bar{1}0]$

FIGURE 5.8

(a) Atom positions in a $(11\bar{2})$ plane perpendicular to a pure edge dislocation lying in the $[11\bar{2}]$ direction and having Burgers vector $\frac{1}{2}[1\bar{1}0]$ in a computer simulation model crystal of copper. The dislocation has dissociated into two Shockley partials at the positions shown. Atom displacements either into or out of the plane of the paper are indicated by smaller or larger circles, respectively. (b) The $(11\bar{2})$ plane viewed at a shallow angle in order to see the edge and screw components of the two partials more clearly. *(Courtesy Yu. N. Osetsky.)*

can be assisted by thermal activation and hence the ease of cross slip decreases with decreasing temperature. Constriction is also assisted by a stress (the *Escaig* stress) acting on the edge component of the two partials so as to push them together.

The sequence of events envisaged during the cross-slip process is illustrated in Fig. 5.10. An extended $\frac{1}{2}\langle 110\rangle$ dislocation, say $\mathbf{b} = \frac{1}{2}[1\bar{1}0]$, lying in the (111) slip plane in (a), has constricted along a short length parallel to the $[1\bar{1}0]$ direction in (b). The constricted dislocation has a pure screw orientation but

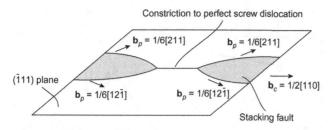

FIGURE 5.9
Constriction to a perfect screw segment in an extended dislocation in a face-centered cubic metal. The Burgers vector of a Shockley partial is denoted by $\mathbf{b}_p$, that of the perfect screw by $\mathbf{b}_c$. *(After Seeger (1957), Dislocations and Mechanical Properties of Crystals, p. 243, Wiley.)*

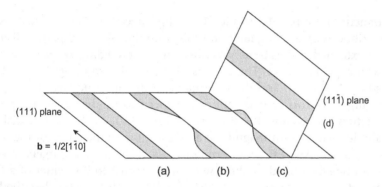

FIGURE 5.10
Four stages in the cross slip of a dissociated dislocation (a) by the formation of a constricted screw segment (b). The screw has dissociated in the cross-slip plane at (c).

is unstable with respect to redissociation. A constriction is likely to form at a region in the crystal, such as a barrier provided by a non-glissile dislocation or impenetrable particle, where the applied stress tends to push the partials together. By stage (c) the unit dislocation has dissociated into two different partial dislocations with a stacking fault but on the $(11\bar{1})$ plane rather than (111). This plane intersects the original glide plane along $[1\bar{1}0]$ and is therefore a possible cross-slip plane. The new extended dislocation is free to glide in the cross-slip plane and has transferred totally to this plane by stage (d).

5.4 THOMPSON'S TETRAHEDRON

Thompson's tetrahedron provides a convenient notation for describing all the important dislocations and dislocation reactions in face-centered cubic metals. It arose from the appreciation that the four different sets of {111} planes lie parallel to the four faces of a regular tetrahedron and the edges of

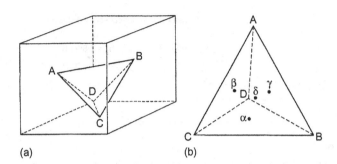

(a)   (b)

FIGURE 5.11
(a) Tetrahedron formed by joining four nearest-neighbor sites *ABCD* in a face-centered cubic structure.
(b) Thompson's tetrahedron.

the tetrahedron are parallel to the ⟨110⟩ slip directions (Fig. 5.11(a)). Note that the ribbons of stacking fault and the bounding Shockley partial dislocations of extended dislocations described in the preceding section are confined to the {111} planes. The corners of the tetrahedron (Fig. 5.11(b)) are denoted by *A*, *B*, *C*, *D*, and the mid-points of the opposite faces by α, β, γ, δ. The Burgers vectors of dislocations are specified by their two end points on the tetrahedron. Thus, the Burgers vectors $\frac{1}{2}\langle 110 \rangle$ of the perfect dislocations are defined both in magnitude and direction by the edges of the tetrahedron and are **AB**, **BC**, etc. Similarly, Shockley partial Burgers vectors $\frac{1}{6}\langle 112 \rangle$ can be represented by the line from the corner to the center of a face, such as **A**β, **A**γ, etc. The dissociation of a $\frac{1}{2}\langle 110 \rangle$ dislocation described by relation (5.2) can be expressed alternatively by reactions of the type:

$$\mathbf{AB} = \mathbf{A}\delta + \delta\mathbf{B} \quad \text{(on slip plane } ABC\text{)}$$
$$\mathbf{AB} = \mathbf{A}\gamma + \gamma\mathbf{B} \quad \text{(on slip plane } ABD\text{)} \tag{5.7}$$

Some caution must be exercised in using this notation for analyzing dislocation reactions, for it is implicit that the two partials only enclose an intrinsic fault when taken in the correct order. For the Burgers circuit construction used here (section 1.4), the rule is as follows. When a perfect dislocation (**b** = **AB**, say) is viewed along the direction of its positive line sense by an observer outside the tetrahedron, an intrinsic fault is produced by a dissociation in which the partial on the left has a Greek-Roman Burgers vector ($\delta\mathbf{B}$ or $\gamma\mathbf{B}$) and that on the right a Roman-Greek one (**A**δ or **A**γ). For an observer inside the tetrahedron, this order is reversed.

5.5 FRANK PARTIAL DISLOCATION

There is an alternative arrangement by which a stacking fault can end in a crystal. The *Frank partial dislocation* is formed as the boundary line of a fault

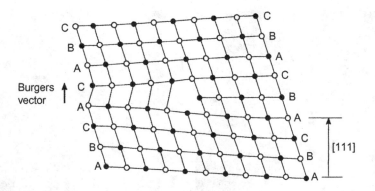

FIGURE 5.12
Formation of a $\frac{1}{3}$[111] Frank partial dislocation by removal of part of a close-packed layer of atoms.
The projection and directions are the same as Fig. 5.2. *(After Read (1953)*, Dislocation in Crystals,
McGraw-Hill.)

formed by inserting or removing one close-packed {111} layer of atoms.
The latter is illustrated in Fig. 5.12. Removal of a layer results in the *intrinsic*
fault with stacking sequence *ABCACABC...* whereas insertion produces the
extrinsic fault with *ABCABACAB...* (See Fig. 1.13.) Geometrically this intrinsic fault is identical to the intrinsic fault produced by dissociation of a perfect dislocation (section 5.3), but the bounding partial is different. The
Frank partial has a Burgers vector *normal* to the {111} plane of the fault and
the magnitude of the vector is equal to the change in spacing produced by
one close-packed layer, i.e. $\mathbf{b} = \frac{1}{3}\langle 111 \rangle$. In Thompson's notation $\mathbf{b} = \mathbf{A}\alpha$ for a
stacking fault in plane *BCD*. The Frank partial is an edge dislocation and
since the Burgers vector is not contained in one of the {111} planes, it cannot glide and move conservatively under the action of an applied stress.
Such a dislocation is said to be *sessile*, unlike the *glissile* Shockley partial.
However, it can move by *climb*.

A closed dislocation loop of a Frank partial dislocation can be produced by
the collapse of a platelet of vacancies as illustrated in Figs 1.13(a) and 3.19:
it may arise from the local supersaturation of vacancies produced by rapid
quenching (section 1.3) or by the displacement cascades formed by irradiation with energetic atomic particles. By convention this is called a *negative
Frank dislocation*. A *positive Frank dislocation* may be formed by the precipitation of a close-packed platelet of interstitial atoms (Figs 1.13(b) and 3.19(d)),
as produced by irradiation damage. Both positive and negative Frank loops
contain stacking faults. Diffraction fringes due to stacking faults (section
2.2) are sometimes observed when dislocation loops are examined in the
electron microscope. An example is given in Fig. 5.13(a); this shows hexagonal vacancy loops formed in an aluminum alloy by quenching. The sides of

FIGURE 5.13

Prismatic and sessile dislocation loops in an aluminum 3.5 per cent magnesium alloy quenched from 550°C into silicone oil at − 20°C. (a) Immediately after quenching some of the loops, e.g. A, contain stacking faults, as seen by fringe contrast, and are Frank sessile dislocations. (b) After being heated slightly the stacking fault in one of the loops has disappeared indicating that the loop is now a perfect dislocation. *(From Westmacott, Barnes, Hull and Smallman,* Phil. Mag. **6***, 929, 1961.)*

the loop are parallel to ⟨110⟩ close-packed directions in the fault plane. In some cases no stacking fault contrast is observed. All the loops initially grow as negative Frank loops nucleated from collapsed vacancy discs on {111} planes, but the stacking fault can be removed by a dislocation reaction as follows. Considering the negative Frank sessile dislocation in Fig. 5.12, the fault will be removed if the crystal above the fault is sheared so that $C \rightarrow B$, $A \rightarrow C$, $B \rightarrow A$, etc. This $\frac{1}{6}\langle 112 \rangle$ displacement corresponds to the glide of a Shockley partial dislocation across the fault. The Shockley partial may have one of three $\frac{1}{6}\langle 112 \rangle$ type vectors lying in the fault plane. It is envisaged that this partial dislocation forms inside the loop and then spreads across the loop removing the fault; at the outside it will react with the Frank partial dislocation to produce a perfect dislocation. One of the three possible reactions for a loop on the (111) plane (or on the *BCD* plane in the Thompson tetrahedron notation) is

$$\frac{1}{6}\left[11\bar{2}\right] + \frac{1}{3}[111] \rightarrow \frac{1}{2}[110]$$

$$\begin{array}{ccc} B\alpha & \alpha A & BA \\ \text{Shockley partial} & \text{Frank partial} & \text{Perfect dislocation} \end{array}$$

(5.8)

Figure 5.13(b) shows the same field as Fig. 5.13(a) after the foil had been in the microscope for some time; the fringe contrast in loop A has disappeared

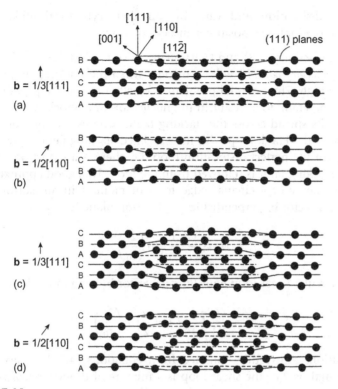

FIGURE 5.14

(a), (c). Atomic structure through vacancy and interstitial Frank loops on the (111) plane of a face-centered-cubic metal, and (b), (d) the perfect loops formed by the unfaulting reactions (5.8), (5.9). *(After Ullmaier and Schilling (1980), in* Physics of Modern Materials, *p. 301, IAEA, Vienna.)*

due to a reaction of the type described above and the Burgers vector of the loop has changed from $\frac{1}{3}\langle 111\rangle$ to $\frac{1}{2}\langle 110\rangle$.

For an interstitial loop, two Shockley partials are required to remove the extrinsic fault. With reference to Fig. 1.13(b), one partial glides below the inserted layer transforming $A \to C$, $C \to B$, $A \to C$, $B \to A$, etc., leaving an intrinsic fault $CABCBCA\ldots$, and the other sweeps above the layer with the same result as in the vacancy case. One possible reaction in, say, the (111) plane is

$$\frac{1}{6}\left[\bar{1}2\bar{1}\right] + \frac{1}{6}\left[2\bar{1}\bar{1}\right] + \frac{1}{3}[111] \to \frac{1}{2}[110]$$

$$\alpha C \qquad\quad \alpha D \qquad\quad \alpha A \qquad\quad BA$$

(5.9)

The atom positions before and after reactions of types (5.8) and (5.9) are shown schematically in Fig. 5.14. The prismatic loops thus formed are rings of perfect dislocation and can slip on their cylindrical glide surfaces (Fig. 3.18) to adopt new positions and orientations.

Unfaulting reactions (5.8) and (5.9) owing to single and double dislocations (S- and D-Shockley partials) will occur only when the stacking fault energy is sufficiently high. The essential problem is whether or not the prevailing conditions in the Frank loop result in the nucleation of a Shockley partial dislocation and its spread across the stacking fault. A necessary condition is that the energy of the Frank loop with its associated stacking fault is greater than the energy of the perfect dislocation loop, i.e. there is a reduction in energy when the stacking fault is removed from the loop. To a good approximation, the elastic energy of a circular edge loop of radius r in an isotropic solid with Burgers vector $\mathbf{b}_e$ perpendicular to the loop plane is

$$E = \frac{Gb_e^2 r}{2(1 - \nu)} \ln\left(\frac{2r}{r_0}\right) \qquad (5.10)$$

and for a circular shear loop with Burgers vector $\mathbf{b}_s$ lying in the loop plane

$$E = \frac{Gb_e^2 r}{2(1 - \nu)} \left(1 - \frac{\nu}{2}\right) \ln\left(\frac{2r}{r_0}\right) \qquad (5.11)$$

These relations are readily obtained from equations (4.22) by noting that the line length is $2\pi r$, the shear loop is a mixture of dislocation of edge and screw character, and the stress fields of dislocation segments on opposite sides of a loop will tend to cancel at distances $\approx 2r$ from the loop, so that the outer cut-off parameter R in (4.22) is $\approx 2r$. For the Frank loop with $\mathbf{b} = \frac{1}{3}\langle 111 \rangle$, $b_e^2 = a^2/3$ and $b_s^2 = 0$, and for the perfect loop with $\mathbf{b} = \frac{1}{2}\langle 110 \rangle$, $b_e^2 = a^2/3$ and $b_s^2 = a^2/6$. Thus, the difference in energy between the Frank loop containing the stacking fault and the perfect, unfaulted loop is

$$\Delta E = \pi r^2 \gamma - \frac{rGa^2}{24}\left(\frac{2 - \nu}{1 - \nu}\right) \ln\left(\frac{2r}{r_0}\right) \qquad (5.12)$$

Therefore, the unfaulting reaction will be energetically favorable if

$$\gamma > \frac{Ga^2}{24\pi r}\left(\frac{2 - \nu}{1 - \nu}\right) \ln\left(\frac{2r}{r_0}\right) \qquad (5.13)$$

Thus the lower limit to the value of γ for removal of a fault depends on the size of the loop. Taking $a = 0.35$ nm, $r = 10$ nm, $G = 40$ GN m^{-2}, $r_0 = 0.5$ nm, and $\nu = 0.33$, the critical stacking fault energy is about 60 mJ m^{-2}. Since $r = 10$ nm is close to the minimum size for resolving loops in the electron microscope it is not surprising that stacking fault loops are rarely observed in metals with γ values larger than this.

The necessary condition based on initial and final energy values may not be a sufficient one for unfaulting, for it neglects the fact that the Shockley partial must be nucleated somewhere within the Frank loop, and there is almost certainly an energy barrier for this process. If a loop grows by the absorption of point defects, it will become increasingly less stable, but if the Shockley nucleation energy is independent of r, the equilibrium unfaulted state may not be achieved. The probability of nucleation is increased by increasing temperature and the presence of external and internal sources of stress. The removal of the fault in Fig. 5.13 may have occurred due to local shear stresses in the foil. Thus, there is no hard-and-fast rule governing loop unfaulting.

Finally, it is noted that if a second platelet of vacancies nucleates within a negative Frank loop adjacent to the intrinsic fault with stacking sequence *ABCACABC...*, a disc of extrinsic fault *ABCACBC...* is formed. If a third platelet nucleates against the second, the stacking sequence through all three is perfect *ABCABC...* Concentric loops of Frank partial dislocation containing alternating rings of intrinsic, extrinsic and perfect stacking have been observed under certain conditions.

5.6 DISLOCATION LOCKS AND STAIR-ROD PARTIALS

Strain hardening in metals can be attributed to the progressive introduction during straining of barriers to the free movement of dislocations (section 10.8). Several barriers have been proposed for the face-centered cubic metals. They form by contact reaction between dislocations of different slip systems. There are three possible $\frac{1}{2}\langle 110 \rangle$ vectors plus their reverses on each of the four $\{111\}$ glide planes (Fig. 5.5), and so 18 reactions are possible between a dislocation of one slip system and dislocations on the other three planes. Reactions are favorable or unfavorable according to Frank's rule for the Burgers vectors involved (section 4.4). If the two dislocations have the same Burgers vector but opposite line sense when they meet along the line of intersection of their glide planes, the reacting segments annihilate, as illustrated by the *collinear interaction* in Fig. 7.18. In other cases, favorable reactions create new dislocations that can act as barriers to other dislocations of the slip systems involved. One is the *Lomer lock*. It can be formed in the following way.

Consider two perfect dislocations gliding in different $\{111\}$ planes (Fig. 5.15(a)). Their Burgers vector and positive line sense are indicated by arrows. If they move to combine along the $[0\bar{1}1]$ intersection line of their

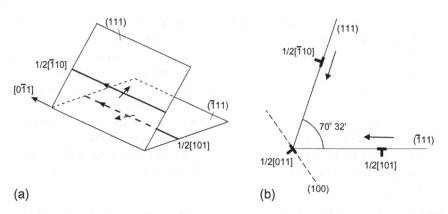

FIGURE 5.15
Formation of a Lomer sessile dislocation. The two $\frac{1}{2}\langle110\rangle\{111\}$ dislocations in (a) can react to form a $\frac{1}{2}\langle110\rangle\{001\}$ dislocation as in (b).

glide planes, as sketched in [0$\bar{1}$1] projection in Fig. 5.15(b), a pure edge dislocation known as a *Lomer dislocation* is formed with Burgers vector given by

$$\tfrac{1}{2}\left[\bar{1}10\right]+\tfrac{1}{2}[101]\rightarrow\tfrac{1}{2}[011] \qquad (5.14)$$

In Thompson tetrahedron notation, **DA** (on ACD) plus **BD** (on BCD) equals **BA**, as illustrated in Fig. 5.16(a). The reaction results in a 50% reduction in dislocation energy per unit length. The glide plane of the Lomer dislocation is (100) and so it is *sessile*, i.e. it cannot glide on any of the four {111} planes.

Another reaction can occur in which b^2 is unchanged but the two reacting dislocations attract to form a stable segment if their initial orientations are suitable. For example, a dislocation on ABD with Burgers vector **DB** (Figs 5.11(a) and 5.16(a)) can react with one on ABC with Burgers vector **AC**. The product vector (**DB** + **AC**) is of <100> type, which is not a slip vector, and the new segment forms an effective lock known as a *Hirth lock*.

In most metals, each dislocation in Fig. 5.15(a) will be dissociated in its glide plane into two Shockley partial dislocations bounding a stacking-fault ribbon (Fig. 5.17(a)). If the dislocations meet at the line of intersection of the two planes, the leading partials repel or attract each other according to Frank's rule for their Burgers vectors. There are three possible $\frac{1}{6}\langle112\rangle$ vectors plus their reverses on each plane (Fig. (5.5)), and there are therefore 36 combinations to consider. The most favorable gives a product dislocation with

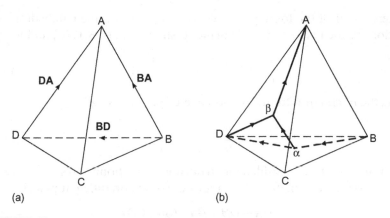

FIGURE 5.16
Thompson tetrahedron illustrating Burgers vectors involved in formation of (a) a Lomer lock and (b) a Lomer—Cottrell lock.

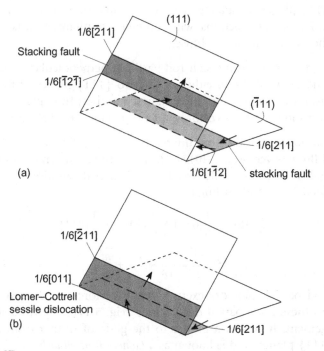

FIGURE 5.17
Formation of a Lomer—Cottrell sessile dislocation. The perfect dislocations in Fig. 5.15(a) have dissociated in (a) and reacted favourably to form a Lomer—Cottrell dislocation in (b).

Burgers vector of the form $\frac{1}{6}\langle 110 \rangle$, i.e. $\alpha\beta$, $\alpha\gamma$, etc., on the tetrahedron. This reaction for the (111) and ($\bar{1}$11) planes is shown in Fig. 5.17(b) and is

$$\frac{1}{6}\left[\bar{1}2\bar{1}\right] + \frac{1}{6}\left[1\bar{1}2\right] \rightarrow \frac{1}{6}[011] \qquad (5.15)$$

Using the b^2 criterion for dislocation energy per unit length:

$$\frac{a^2}{6} + \frac{a^2}{6} \rightarrow \frac{a^2}{18} \qquad (5.16)$$

which represents a considerable reduction. In Thompson's notation for Burgers vectors, two perfect dislocations dissociate on different planes:

$$\begin{aligned} \mathbf{DA} &\rightarrow \mathbf{D}\beta + \beta\mathbf{A} \quad \text{(on ACD)} \\ \mathbf{BD} &\rightarrow \mathbf{B}\alpha + \alpha\mathbf{D} \quad \text{(on BCD)} \end{aligned} \qquad (5.17)$$

and one Shockley from each combine:

$$\alpha\mathbf{D} + \mathbf{D}\beta \rightarrow \alpha\beta \qquad (5.18)$$

The vectors are shown in Fig. 5.16(b). The product partial dislocation forms along one of the six $\langle 110 \rangle$ directions at the intersection of the stacking faults on two {111} planes. By analogy with carpet on a stair, it is called a *stair-rod dislocation*. For some reactions with different resultant vectors, the angle between the stacking faults is obtuse.

The Burgers vector $\alpha\beta$ of the stair-rod partial is perpendicular to the dislocation line and does not lie in either of the two {111} planes of the adjacent faults. Thus, it cannot glide on these planes, and the {100} plane which contains the line and its Burgers vector is not a slip plane. The dislocation is *sessile*.

The dislocations that result from the reactions in Figs 5.15 and 5.17 have the same total Burgers vector, i.e. $\frac{1}{2}[011]$. However, the arrangement of three partials in Fig. 5.17 is more stable because the Lomer dislocation can dissociate with a reduction in b^2 according to

$$\frac{1}{2}[011] \rightarrow \frac{1}{6}[011] + \frac{1}{6}\left[\bar{2}11\right] + \frac{1}{6}[211] \qquad (5.19)$$

$$b^2 : \frac{a^2}{2} \rightarrow \frac{a^2}{18} + \frac{a^2}{6} + \frac{a^2}{6}$$

The stair-rod partial therefore exerts a repulsive force on the two Shockley partials and these three partial dislocations (Fig. 5.17(b)) form a stable, sessile arrangement. It acts as a barrier to the glide of further dislocations on the two {111} planes, and is known as a *Lomer–Cottrell lock*.

Stair-rod dislocations described above are the reaction product of dislocations from different slip systems. Stair-rod partials are also geometrically

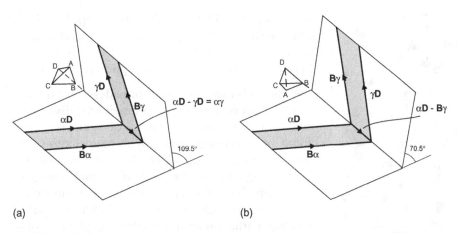

FIGURE 5.18
Dissociated dislocation with total Burgers vector BD bending from plane BCD to plane ADB through
(a) an acute angle and (b) an obtuse angle. A stair-rod partial is created at the bend.

necessary when a dissociated dislocation bends from one plane to another.
Figure 5.18 uses the Thompson tetrahedron notation for a dissociated dislocation bending from plane BCD to plane ABD. Its total Burgers vector is **BD**,
which lies in both planes. The angle between the two planes may be acute,
as in (a), or obtuse, as in (b). The arrows on the partial dislocations show
the positive line sense used to define the Burgers vectors according to the tetrahedron drawn on each figure. (Following the rule stated at the end of section 5.4, Shockley partials on plane ABD in Fig. 5.18(b) have Greek letters
on the inside, consistent with an observer viewing from inside the tetrahedron.) The Burgers vectors of the Shockley partials change with the change
of plane, and so the rule that Burgers vector be conserved (section 1.4)
requires a stair-rod partial to exist at each bend. Application of the rule at
either of the nodes at the ends of the stair-rods shows that the Burgers vector
of the stair-rod is $\alpha\gamma$ in the acute case, i.e. $(\alpha D + D\gamma)$ or $(\alpha B + B\gamma)$, and
$(\alpha D + \gamma B)$ or $(\alpha B + \gamma D)$ for the obtuse case. These have Miller indices of
the type $\frac{1}{6}\langle 110 \rangle$ and $\frac{1}{3}\langle 100 \rangle$, respectively.

5.7 STACKING FAULT TETRAHEDRA

Another dislocation arrangement has been observed in metals and alloys of
low stacking-fault energy following treatment that produces a supersaturation
of vacancies. It consists of a tetrahedron of intrinsic stacking faults on {111}
planes with $\frac{1}{6}\langle 110 \rangle$ type stair-rod dislocations along the edges of the tetrahedron. Stacking-fault tetrahedra have been observed in metals following

FIGURE 5.19

Transmission electron micrograph of tetrahedral defects in quenched gold. The shape of the tetrahedra viewed in transmission depends on their orientation with respect to the plane of the foil, (110) foil orientation. *(From Cottrell, Phil. Mag. **6**, 1351, 1961.)*

FIGURE 5.20

Jog lines on a {111} face of a stacking-fault tetrahedron.

quenching (Fig. 5.19) or radiation damage (Fig. 2.9). Once nucleated, they can grow in a supersaturation of vacancies by the climb of ledges ('*jog lines*') on the {111} faces due to vacancy absorption (Fig. 5.20).

The *Silcox–Hirsch mechanism* provides one explanation for their occurrence. According to the discussion in section 5.5, when a platelet of vacancies (produced by quenching from a high temperature) collapses to form a loop of Frank partial dislocation, the stacking fault will be stable if the fault energy is sufficiently low. The Frank partial may dissociate into a low-energy stair-rod dislocation and a Shockley partial on an intersecting slip plane according to a reaction of the type

$$\frac{1}{3}[111] \rightarrow \frac{1}{6}[101] + \frac{1}{6}[121]$$

$$b^2 : \frac{a^2}{3} \rightarrow \frac{a^2}{18} + \frac{a^2}{6} \tag{5.20}$$

Discounting the energy of the stacking fault there is a reduction in b^2 and the reaction is favorable according to Frank's rule. With reference to the Thompson tetrahedron notation (Fig. 5.11(b)), suppose that the vacancies condense on the {111} plane *BCD* in the form of an equilateral triangle with edges parallel to ⟨110⟩ directions *BC*, *CD*, *DB*. The triangular Frank partial with Burgers vector αA can dissociate to produce a stair-rod along each edge and a Shockley partial on each of the three inclined {111} planes as shown in Fig. 5.21(a) by Burgers vector reactions of the type (5.20), namely

$$\begin{aligned} \alpha A &\rightarrow \alpha \beta + \beta A \quad \text{(on ACD)} \\ \alpha A &\rightarrow \alpha \gamma + \gamma A \quad \text{(on ABD)} \\ \alpha A &\rightarrow \alpha \delta + \delta A \quad \text{(on ABC)} \end{aligned} \tag{5.21}$$

The Shockley partial dislocations βA, γA and δA will be repelled by the stair-rod dislocations $\alpha\beta$, $\alpha\gamma$ and $\alpha\delta$ respectively and will bow out in their slip planes. Taking account of dislocation line sense, it is found that the partials attract each other in pairs to form another set of stair-rods along *DA*, *BA* and *CA* (Fig. 5.21(b)) according to the reactions

$$\begin{aligned} \beta A + A\gamma &\rightarrow \beta\gamma \quad \text{(along DA)} \\ \gamma A + A\delta &\rightarrow \gamma\delta \quad \text{(along BA)} \\ \delta A + A\beta &\rightarrow \delta\beta \quad \text{(along CA)} \end{aligned} \tag{5.22}$$

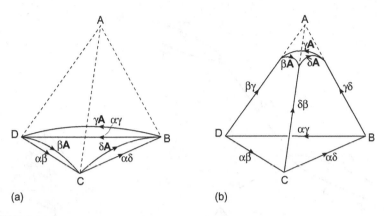

FIGURE 5.21

Formation of a stacking-fault tetrahedron by the Silcox–Hirsch mechanism. Arrows show the positive line sense used to define the Burgers vectors, which are denoted by directions on the Thompson tetrahedron.

In Miller index notation the reactions are of the type (5.15). As a result of the transformation to a stacking fault tetrahedron, the vacancy content of the original Frank loop becomes distributed equally over the four faces of the tetrahedron.

The shape of tetrahedra observed in thin foils by transmission microscopy depends on the orientation of the tetrahedra with respect to the plane of the foil, as seen in Fig. 5.19. The complex contrast patterns inside the faults arise from overlapping stacking faults in different faces of the tetrahedron. The increase in energy due to the formation of stacking faults places a limit on the size of the fault that can be formed. If the fault energy is relatively high, the Frank loop may be stable or it may only partly dissociate, thereby forming a truncated tetrahedron as in Fig. 5.21(b).

The stacking fault tetrahedra observed in irradiated metals (Fig. 2.9) are smaller than those created as a result of quenching (Fig. 5.19). They are believed to form directly in displacement cascades, rather than by the Silcox–Hirsch mechanism. The core region of a cascade is disordered with a high concentration of vacancies and molecular dynamics computer simulation shows that a tetrahedron can form as the atoms reorganize to a more stable arrangement. Stacking fault tetrahedra can also result from plastic deformation due to the cross slip of a segment of jogged screw dislocation, as illustrated in Fig. 5.22. In (a), the screw dislocation with Burgers vector **AD** (Thompson tetrahedron notation) is dissociated in the plane *ABD* (**AD** = Aγ + γD) and contains a jog. The right-hand segment of the screw

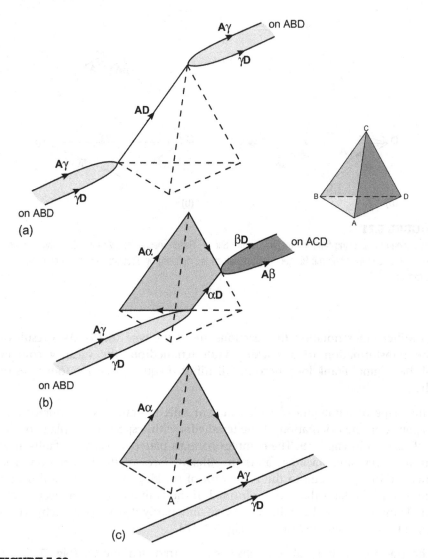

FIGURE 5.22

Three stages in the formation of a triangular Frank loop by cross slip of a jogged screw dislocation. Pairs of letters denote Burgers vectors in the notation of Fig. 5.11.

has cross slipped in (b) onto plane ACD ($AD = A\beta + \beta D$) and the jog has dissociated on plane BCD into a Frank partial ($A\alpha$) and a Shockley partial (αD). The jog has been removed in (c): the screw dislocation glides away leaving a triangular Frank loop (Burgers vector $A\alpha$) on BCD which can form a stacking-fault tetrahedron by the Silcox–Hirsch mechanism of Fig. 5.21.

FURTHER READING

Amelinckx S: Dislocations in particular structures. In Nabarro FRN, editor: *Dislocations in solids*, vol 2, 1979, North-Holland, p. 66.

Cottrell AH: *Dislocations and plastic flow in crystals*, 1953, Oxford University Press.

Eyre BL, Loretto MH, Smallman RE: Electron microscopy studies of point defect clusters in metals. In Smallman RE, Harris JE, editors: *Vacancies '76*, London, 1977, The Metals Society, p. 63.

Hirth JP, Lothe J: *Theory of dislocations*, 1982, Wiley.

Kelly A, Groves GW, Kidd P: *Crystallography and crystal defects*, 2000, John Wiley.

Read WT: *Dislocations in crystals*, New York, 1953, McGraw-Hill.

Thompson N: Dislocation nodes in face-centered cubic lattices, *Proc Phys Soc B* 66:481, 1953.

Veysièrre P: Dislocations and the plasticity of crystals. In Meyers MA, Armstrong RW, Kirchner H, editors: *Mechanics and materials: fundamental linkages*, 1999, John Wiley, p. 271.

Whelan MJ: Dislocation interactions in face-centered cubic metals, *Proc Roy Soc A* 249:114, 1959.

Dislocations in Other Crystal Structures

6.1 INTRODUCTION

The face-centered cubic metals have been treated separately in Chapter 5 because although many of the dislocation reactions and properties presented have counterparts in other structures, they are more readily described in the face-centered cubic system. In general, reducing the crystal symmetry, changing the nature of the interatomic bonding and increasing the number of atom species in the lattice make dislocation behavior more complex. Nevertheless, many of the features of the preceding chapter carry over to other structures, as will be seen in the following. The two other major metallic structures are discussed first, and then some important compounds and non-metallic cases are considered.

6.2 DISLOCATIONS IN HEXAGONAL CLOSE-PACKED METALS

Burgers Vectors and Stacking Faults

Some of the important metals of this structure are given in Table 6.1. As explained in section 1.2, the (0001) basal plane is close-packed and the close-packed directions are $\langle 11\bar{2}0 \rangle$. The shortest lattice vectors are $\frac{1}{3}\langle 11\bar{2}0 \rangle$, the unit cell generation vectors **a** in the basal plane. It may be anticipated, therefore, that dislocation glide will occur in the basal plane with Burgers vector $\mathbf{b} = \frac{1}{3}\langle 11\bar{2}0 \rangle$: this slip system is frequently observed. None of the metals has a structure represented by ideally close-packed atomic spheres, which would require the lattice parameter ratio c/a to be $(8/3)^{1/2} = 1.633$, although magnesium and cobalt have a c/a ratio close to ideal. This indicates that directionality occurs in interatom bonding. As a consequence of this, it is found that some metals slip most easily with $\mathbf{b} = \frac{1}{3}\langle 11\bar{2}0 \rangle$ on the first-order prism planes $\{10\bar{1}0\}$ (see Table 6.1). Basal and prism planes, and a

109

Table 6.1 Properties of Some Hexagonal Close-Packed Metals at 300 K

Metal	Be	Ti	Zr	Mg	Co	Zn	Cd
c/a ratio	1.568	1.587	1.593	1.623	1.628	1.856	1.886
Preferred slip plane for **b = a**	basal (0001)	prism {10$\bar{1}$0}	prism {10$\bar{1}$0}	basal (0001)	basal (0001)	basal (0001)	basal (0001)

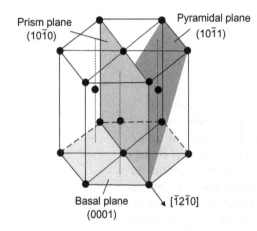

Prism plane (10$\bar{1}$0)   Pyramidal plane (10$\bar{1}$1)

Basal plane (0001)   [$\bar{1}$2$\bar{1}$0]

FIGURE 6.1

Planes in an hexagonal lattice with common [$\bar{1}$2$\bar{1}$0] direction.

first-order pyramidal plane with a common [$\bar{1}$2$\bar{1}$0] axis are shown in Fig. 6.1.

Burgers vectors for the structure may be described in a similar fashion to the Thompson tetrahedron for face-centered cubic metals by using the bi-pyramid shown in Fig. 6.2. The important dislocations and their Burgers vectors are as follows.

(a) Perfect dislocations with one of six Burgers vectors in the basal plane along the sides of the triangular base *ABC* of the pyramid, represented by **AB, BC, CA, BA, CB** and **AC**.

(b) Perfect dislocations with one of two Burgers vectors perpendicular to the basal plane, represented by the vectors **ST** and **TS**.

(c) Perfect dislocations with one of twelve Burgers vectors represented by symbols such as **SA/TB**, which means either the sum of the vectors **ST** and **AB** or, geometrically, a vector equal to twice the join of the midpoints of *SA* and *TB*.

(d) Imperfect basal dislocations of the Shockley partial type with Burgers vectors **Aσ, Bσ, Cσ, σA, σB** and **σC**.

(e) Imperfect dislocations with Burgers vectors perpendicular to the basal plane, namely, **σS, σT, Sσ** and **Tσ**.

(f) Imperfect dislocations which are a combination of the latter two types given by **AS, BS**, etc. Although these vectors represent a displacement from one atomic site to another the associated dislocations are not perfect. This is because the sites do not have identical surroundings and the vectors are not translations of the lattice.

The Miller–Bravais indices and length b of these Burgers vectors **b** are given in Table 6.2. The value of b^2 for ideal close-packing ($c^2 = \frac{8}{3}a^2$) is also given: appropriate adjustments are required when dealing with non-ideal packing.

A number of different stacking faults are associated with the partial dislocations listed in Table 6.2. According to the hard-sphere model of atoms, three basal-plane faults exist which do not affect nearest-neighbor arrangements of the perfect stacking sequence $ABABAB$... Two are intrinsic and conventionally called I_1 and I_2. Fault I_1 is formed by removal of a basal layer, which produces a very high energy fault, followed by slip of $\frac{1}{3}\langle 10\overline{1}0 \rangle$ of the crystal above this fault to reduce the energy:

$$ABABABABA\ldots \rightarrow ABABBABA\ldots \rightarrow ABABCBCB\ldots (I_1) \qquad (6.1)$$

Fault I_2 results from slip of $\frac{1}{3}\langle 10\overline{1}0 \rangle$ in a perfect crystal:

$$ABABABAB\ldots \rightarrow ABABCACA\ldots (I_2) \qquad (6.2)$$

The extrinsic fault (E) is produced by inserting an extra plane:

$$ABABABAB\ldots \rightarrow ABABCABAB\ldots (E) \qquad (6.3)$$

FIGURE 6.2
Burgers vectors in the hexagonal close-packed lattice. *(From Berghezan, Fourdeux and Amelinckx, Acta Metall. 9, 464, 1961.)*

Table 6.2 Dislocations in Hexagonal Close-Packed Structures. The Values of b^2 are for the Case $c^2 = \frac{8}{3}a^2$

Type	AB	TS	SA/TB	Aσ	σS	AS
b	$\frac{1}{3}\langle 11\overline{2}0 \rangle$	$[0001]$	$\frac{1}{3}\langle 11\overline{2}3 \rangle$	$\frac{1}{3}\langle \overline{1}100 \rangle$	$\frac{1}{2}[0001]$	$\frac{1}{6}\langle \overline{2}203 \rangle$
b	a	c	$(c^2 + a^2)^{\frac{1}{2}}$	$a/\sqrt{3}$	$c/2$	$\left(\frac{a^2}{3} + \frac{c^2}{4} \right)^{\frac{1}{2}}$
b^2	a^2	$\frac{8}{3}a^2$	$\frac{11}{3}a^2$	$\frac{1}{3}a^2$	$\frac{2}{3}a^2$	a^2

These faults introduce into the crystal a thin layer of face-centered cubic stacking (ABC) and so have a characteristic stacking-fault energy γ. The main contribution to γ arises from changes in the second-neighbor sequences of the planes. There is one change for I_1, i.e. the A sequence changes to C, two for I_2 and three for E, and so to a first approximation $\gamma_E \approx \frac{3}{2}\gamma_{I_2} \approx 3\gamma_{I_1}$. Experimental estimates of γ for the basal-slip metals are generally higher than those quoted earlier for the face-centered cubic metals. However, theoretical values obtained by *ab initio* methods (section 2.4) are now considered to be more reliable. Values for γ for the I_2 basal fault in magnesium are in the range 30 to 40 mJm^{-2} and calculations for zirconium give 200 mJm^{-2}.

The symmetry of the face-centered cubic structure guarantees that the $\{111\}$ γ-surface has an extremum (maximum or minimum) for the fault vector $\frac{1}{6}\langle 11\bar{2}\rangle$ because three mirror planes perpendicular to the fault plane exist at that position (see Figs 5.3 and 5.4). The same condition applies to the I_2 fault on the basal plane of the hexagonal close-packed metals. It is not satisfied for the first-order prism planes $\{10\bar{1}0\}$ in these metals, however.

Geometrical models used to consider possible faults on the $\{10\bar{1}0\}$ planes have led to proposals that stable faults may exist with vectors $\frac{1}{18}\langle 246\bar{3}\rangle$ or $\frac{1}{9}\langle 11\bar{2}0\rangle$. The first produces a fault which is stable in a lattice consisting of hard spheres. The second results in prism planes adjacent to the fault ribbon adopting the stacking of $\{112\}$ planes in body-centered cubic metals, the relevance being that this is the stable crystal structure for titanium and zirconium at high temperature. There is no evidence that these faults exist in real metals, however. Treatment based on *ab initio* calculations for zirconium have shown that a stable fault exists with vector $\frac{1}{6}\langle 11\bar{2}\alpha\rangle$, where α is between 0 and 0.6. The corresponding value of γ is approximately 140 mJm^{-2}. Prism-plane faults in metals such as magnesium that slip predominantly on the basal plane are believed to have very high energy and are probably unstable.

Basal and Non-basal Slip

The shortest lattice vectors of the hexagonal close-packed metals are of the form $\frac{1}{3}\langle 11\bar{2}0\rangle$ (see Table 6.2) and the most commonly observed slip systems are $\frac{1}{3}\langle 11\bar{2}0\rangle(0001)$ and $\frac{1}{3}\langle 11\bar{2}0\rangle\{1\bar{1}00\}$ (see Table 6.1). For metals such as beryllium, magnesium, cadmium and zinc, the critical resolved shear stress (CRSS) for basal slip is low ($\lesssim 1$ MN m^{-2}), whereas for titanium and zirconium, in which prism-plane slip, the critical stress for prism slip is higher ($\gtrsim 10$ MN m^{-2}). Intuitively, the higher value arises from the corrugated nature of the $\{10\bar{1}0\}$atomic planes. However, the choice of preferred glide plane is less easy to explain. Although it may be anticipated that the lattice

FIGURE 6.3
Edge and screw dislocations with $b = \frac{1}{3}\langle 11\bar{2}0 \rangle$ responsible for basal and prism slip in the hexagonal close-packed structure.

resistance to glide is smaller for planes with wide spacing (section 10.3), this does not account for the choice of preferred slip system in some of the hexagonal metals because the inter-planar spacing is $c/2$ for (0001) and $\sqrt{3}a/2$ for $\{10\bar{1}0\}$, and therefore larger for the prism planes when $c/a < \sqrt{3}$. It is now understood that the preference for slip by one system over another is determined by the energy and stability of stacking faults on the basal and prism planes, and by the effect such faults have on the atomic structure in the core of the screw dislocation in particular. The edge and screw dislocations with $b = \frac{1}{3}\langle 11\bar{2}0 \rangle$ that are responsible for basal and prism slip are shown schematically in Fig. 6.3.

If the stable I_2 stacking fault with vector $\frac{1}{3}\langle 11\bar{2}0 \rangle$ described above exists on the basal plane, then $\frac{1}{3}\langle 11\bar{2}0 \rangle(0001)$ slip is similar to $\frac{1}{2}\langle 1\bar{1}0 \rangle\{111\}$ slip in the face-centered cubic metals, in that the critical resolved shear stress (CRSS) is low ($\lesssim 1$ MN m^{-2}) and the perfect dislocation dissociates into two Shockley partials bounding a ribbon of stacking fault. The Burgers vector reaction is

$$\mathbf{AB} \rightarrow \mathbf{A\sigma} + \mathbf{\sigma B} \tag{6.4}$$

e.g.

$$\frac{1}{3}[11\bar{2}0] \rightarrow \frac{1}{3}[10\bar{1}0] + \frac{1}{3}[01\bar{1}0]$$

$$b^2: \qquad a^2 \qquad a^2/3 \qquad a^2/3$$

The geometry is the same as that of the face-centered cubic case, in that the partial vectors lie at $\pm 30°$ to the perfect vector and the fractional reduction in dislocation energy given by b^2 is 1/3. The spacing of the partials may be calculated from equations (5.4) and (5.5) by replacing b^2 by $a^2/3$. The fault

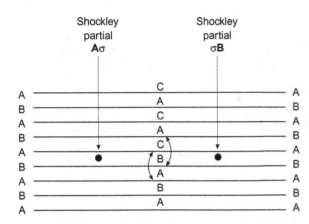

FIGURE 6.4

Dissociation of a perfect dislocation with Burgers vector **AB** (Fig. 6.2) into two Shockley partial dislocations separated by a stacking fault I_2. Double arrows indicate the two errors in the two-fold stacking sequence of the basal planes.

involved is the intrinsic fault I_2, as shown schematically in Fig. 6.4. With the dissociation described, the atomic disregistry of the edge and screw dislocations of Figs 6.3(a) and (c) will be spread along the basal plane. This reduces the critical resolved shear stress required for glide on that plane.

If the stable stacking fault with vector $\frac{1}{6}\langle 11\bar{2}\alpha \rangle$ described above exists on the prism plane, then slip will be similar to that on the basal plane except that the atomic misfit in the core of the dislocations depicted in Figs 6.3(b) and (c) will spread on the prism plane. Again, the spacing of the partial dislocations involved, in this case with Burgers vectors $\frac{1}{6}\langle 11\bar{2}\alpha \rangle$ and $\frac{1}{6}\langle 11\bar{2}\bar{\alpha} \rangle$, is inversely proportional to γ of the fault, and dissociation reduces the critical resolved shear stress for glide on the fault plane.

The significance of having different stacking faults on the basal and prism planes is as follows. Edge dislocations (Fig. 6.3(a) and (b)) can dissociate into partial dislocations separated by a ribbon of fault on either of these planes if stable faults exist. Atomic-scale computer simulation indicates that the CRSS for glide will be low, particularly for the basal system. The screw dislocation (Fig. 6.3(c)) can dissociate on either plane, but the relative values of the stacking fault energy will dictate which plane is chosen. Thus, in magnesium, which as noted above has a stable I_2 basal fault with $\gamma = 30-40$ mJ m^{-2} and a prism-plane fault, possibly unstable, with much higher energy, the screw dislocation in its lowest energy configuration dissociates on the basal plane. It would only be able to glide on the prism plane by first constricting, as explained for the face-centered cubic structure in section 5.3. The same situation is believed to exist in other basal-slip metals such as beryllium, cadmium and zinc. In zirconium, on the other hand, *ab initio* calculations indicate that γ for the basal and prism faults is 200 and 140 mJ m^{-2}, respectively, and so the screw dislocation dissociates preferentially on the prism plane and would have to constrict in order to cross-slip onto the basal plane. The screw dislocation in titanium is believed to have a similar form.

In polycrystalline metals, basal and prism slip do not supply sufficient slip modes to satisfy von Mises' criterion that every grain should be able to plastically deform generally to meet the shape changes imposed by its neighbors (section 10.9). This requires five independent slip systems, whereas basal and prism slip provide only two each. Consequently, twinning and occasionally other slip systems play an important role in the plasticity of these metals.

A variety of twin modes with different habit plane and shear direction occur, depending on the metal, temperature and nature of loading. A deformation twin grows or shrinks under a resolved shear stress by the motion of steps along its boundary. Such steps have dislocation character and are known as *twinning dislocations*. They are described in more detail in section 9.7.

Slip with Burgers vector $a + c = \frac{1}{3}\langle 11\bar{2}3 \rangle$ has been widely reported, although only under conditions of high stress and orientations in which more favored slip vectors cannot operate. The glide planes are $\{10\bar{1}1\}$ (Fig. 6.1) and $\{11\bar{2}2\}$. The magnitude of the Burgers vector is large and the planes are atomically rough, which explains the high glide stress. It has been proposed on the basis of the hard-sphere lattice model that the dislocation on $\{11\bar{2}2\}$ may dissociate into four partials with approximately equal Burgers vectors. One has a core spread over three successive $\{11\bar{2}2\}$ planes, within which the atomic displacements are different. These three adjacent faults are bounded by the three remaining partials. Significantly, the first dislocation is a *zonal* twinning dislocation, which is a multi-layer step in a coherent twin boundary which generates by its movement a twin. Thus, there may be a close relationship between $\langle 11\bar{2}3 \rangle \{11\bar{2}2\}$ slip and twinning, although not necessarily of the hard-sphere lattice form.

Vacancy and Interstitial Loops

As in the face-centered cubic metals, vacancies and interstitials in excess of the equilibrium concentration can precipitate as platelets to form dislocation loops. The situation is more complicated, however, because the relative density of atoms in the different crystallographic planes varies with c/a ratio and stacking faults occur in some metals and not others, as discussed in the preceding section. The simplest geometries are considered here, starting with basal-plane loops.

Condensation of vacancies in a single basal plane (Fig. 6.5(a)) results in two similar atomic layers coming into contact (Fig. 6.5(b)). This unstable situation of high energy is avoided in one of two ways. In one, the stacking of one layer adjacent to the fault is changed; for example B to C as in Fig. 6.5 (c). This is equivalent to the glide below the layer of one Shockley partial with Burgers vector $\frac{1}{3}\langle 10\bar{1}0 \rangle$ followed by glide above of a second Shockley of opposite sign. The Burgers vector reaction is

$$\sigma S + \sigma A + A\sigma \rightarrow \sigma S$$

e.g.

$$\frac{1}{2}[0001] + \frac{1}{3}[1\bar{1}00] + \frac{1}{3}[\bar{1}100] \rightarrow \frac{1}{2}[0001] \qquad (6.5)$$

FIGURE 6.5

Formation of prismatic dislocation loops as a result of the precipitation of one layer of vacancies. (a) Disc-shaped cavity. (b) Collapse of the disc bringing two B layers together. (c) Formation of a high-energy stacking fault E. (d) Formation of a low-energy stacking fault I_1. The actual sequence of planes formed will depend on the plane in which the vacancies form, i.e. A or B. It is possible to have loops of the same form with different Burgers vectors lying on adjacent planes. *(After Berghezan, Fourdeux and Amelinckx,* Acta Metall. *9, 464, 1961.)*

The resultant sessile Frank partial with $b^2 = c^2/4 \simeq 2a^2/3$ surrounds the extrinsic (E) stacking fault described earlier in this section. In the alternative mechanism, a single Shockley partial sweeps over the vacancy platelet, displacing the atoms above by $\frac{1}{3}\langle 10\bar{1}0\rangle$ relative to those below. The Burgers vector reaction is

$$A\sigma + \sigma S \to AS$$

e.g.

$$\frac{1}{3}[\bar{1}100] + \frac{1}{2}[0001] \to \frac{1}{6}[\bar{2}203] \tag{6.6}$$

This sessile Frank partial with $b^2 \simeq a^2$ surrounds the type I_1 intrinsic fault (Fig. 6.5(d)). The E-type loop of reaction (6.5) can transform to the I_1 form by reaction (6.6), and since γ_E is expected to be approximately three times γ_{I_1}, the $\frac{1}{6}\langle 20\bar{2}3\rangle$ loops may be expected to dominate. The dislocation energy is proportional to b^2, however, and the total energy change accompanying reaction (6.6) is dependent on the γ values and loop size, in a similar manner to the unfaulting of loops in face-centered cubic metals (section 5.5).

There is therefore a critical loop size for the reaction, which may be influenced by factors such as stress, temperature and impurity content. Experimentally, both forms are observed in quenched or irradiated metals.

Precipitation of a basal layer of interstitials as in Figs 6.6(a) and (b) produces an E-type fault surrounded by a Frank partial loop with Burgers vector $\frac{1}{2}[0001]$. Again, provided the loop is large enough, this can transform to an I_1 loop by the nucleation and sweep of a Shockley partial according to reaction (6.6). The stacking sequence is shown in Fig. 6.6(c). Interstitial loops with Burgers vectors $\frac{1}{2}[0001]$ and $\frac{1}{6}\langle 20\overline{2}3\rangle$ have been seen in irradiated magnesium, cadmium and zinc. In the latter two metals, perfect loops with Burgers vector [0001] are also observed. They result from a double layer of interstitials, and as they grow during irradiation, some loops transform into two concentric $\frac{1}{2}[0001]$ loops, as shown in Fig. 6.7.

FIGURE 6.6
(a) Precipitation of a layer of interstitials. (b) Prismatic loop resulting from the layer of interstitials: the loop contains a high-energy stacking fault E. (c) Prismatic loop containing low-energy stacking fault I_1. (After Berghezan, Fourdeux and Amelinckx, Acta Metall. *9, 464, 1961.*)

The atomic density of the basal planes is only greater than that of the corrugated $\{10\overline{1}0\}$ prism planes when $c/a > \sqrt{3}$, suggesting that the existence of basal stacking faults with relatively low γ aids the stability of the basal vacancy and interstitial loops in magnesium. These faults are not believed to occur in titanium and zirconium, however, and large basal-plane loops are not expected. This is confirmed experimentally by transmission-electron microscopy of irradiated specimens, which reveals that basal-plane loops with a **c** component in their Burgers vector occur only rarely, for example at high levels of radiation damage. Generally, both vacancy and interstitial loops with the perfect Burgers vector $\frac{1}{3}\langle 11\overline{2}0\rangle$ are seen. They lie on planes of the [0001] zone at angles up to $30°$ from the pure-edge $\{11\overline{2}0\}$ orientation. The $\{11\overline{2}0\}$ planes are neither widely-spaced nor densely-packed, and it is possible that the point defects precipitate initially as single-layer loops on the $\{10\overline{1}0\}$ prism planes. The resulting stacking fault has a relatively high energy and is removed by shear when the loops are small by the Burgers vector reaction

$$\frac{1}{2}\langle 10\overline{1}0\rangle + \frac{1}{6}\langle \overline{1}2\overline{1}0\rangle \rightarrow \frac{1}{3}\langle 11\overline{2}0\rangle \qquad (6.7)$$

FIGURE 6.7

Transmission electron micrographs of interstitial loops with **b** = [0001] in zinc. (a) Below the threshold voltage at which electrons displace zinc atoms and (b) above. Loops β split into concentric loops γ as they grow. *(From Eyre, Loretto and Smallman (1976), Proc. Vacancies '76, p. 63, The Metals Society, London.)*

The glissile loops thus produced can adopt the variety of orientations observed in practice. The small loops shown in Fig. 2.8 are examples of vacancy loops with $\mathbf{b} = \frac{1}{2}\langle 10\bar{1}0 \rangle$ or $\frac{1}{3}\langle 11\bar{2}0 \rangle$ in ruthenium irradiated with heavy ions.

6.3 DISLOCATIONS IN BODY-CENTERED CUBIC METALS

In body-centered cubic metals (e.g. iron, molybdenum, tantalum, vanadium, chromium, tungsten, niobium, sodium and potassium) *slip* occurs in close-packed ⟨111⟩ directions. The shortest lattice vector, i.e. the Burgers vector of the perfect slip dislocation, is of the type $\frac{1}{2}\langle 111 \rangle$. The crystallographic slip planes are {110}, {112} and {123}. Each of these planes contains ⟨111⟩ slip directions and it is particularly significant that three {110}, three {112} and six {123} planes intersect along the same ⟨111⟩ direction. Thus, if cross slip is easy it is possible for screw dislocations to move in a haphazard way on different {110} planes or combinations of {110} and {112} planes, etc., favored by the applied stress. For this reason slip lines are often wavy and ill-defined (see Fig. 3.10). It has been found that the apparent slip plane varies with composition, crystal orientation, temperature and strain rate. Thus, when pure iron is deformed at room temperature the slip plane appears to be close to the maximum resolved shear stress plane irrespective of the orientation, whereas when it is deformed at low temperatures, or alloyed with silicon, slip tends to be restricted to a specific {110} plane.

An interesting feature of yielding is the *asymmetry of slip*. It is found, for example, that the slip plane of a single crystal deformed in uniaxial compression may be different from the slip plane which operates in tension for the same crystal orientation. In other words, the shear stress to move a dislocation in one direction in a slip plane is not the same as the shear stress required to move it in the opposite direction in the same plane. Slip is easier when the applied stress is such that a dislocation would move in the twinning sense (see below) on {112} planes rather than the anti-twinning sense, even when the actual slip plane is not {112}. Electron microscopy of metals deformed at low temperature reveals long screw dislocations, implying that non-screw dislocations are more mobile and that, as in the hexagonal close-packed metals in section 6.2, screw dislocations dictate the slip characteristics.

Stable stacking faults have not been observed experimentally in the body-centered cubic metals, and *ab initio* calculations and atomic-scale modeling using interatomic potentials for these metals (section 2.4) have not revealed minima corresponding to stable faults on the {110} and {112} γ-surfaces. This is consistent with the ease of cross-slip. As in the hexagonal close-packed metals (section 6.2), slip behavior in the body-centered cubic metals is determined by the atomic structure of the core region of the *screw* dislocation. Unlike the structure in the hexagonal case, however, the screw core in the body-centered cubic metals has a distinctive non-planar character. This leads to high lattice resistance to glide of the screw dislocation and consequently to strong temperature and strain-rate dependencies of the yield stress, and to the other features described in the preceding paragraph.

FIGURE 6.8

Orientation of the {110} and {112} planes of the [111] zone and directions of their normals.

The core structure is determined in part by the crystal symmetry. The relevant symmetry for the screw dislocation with $b = \frac{1}{2}[111]$ is due to the fact that the [111] direction is a three-fold screw axis, i.e. three-fold rotation, as demonstrated by the traces of the {110} and {112} planes of the [111] zone in Fig. 6.8, plus a $\pm\frac{1}{6}[111]$ displacement. The atomic structure of the screw dislocation must exhibit this symmetry. Note also that the [10$\bar{1}$] direction normal to the (10$\bar{1}$) plane is a two-fold axis. Atomic-scale computer simulation using interatomic potentials for models of a variety of body-centered cubic metals has revealed two different forms of core structure

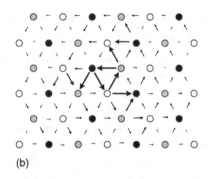

(a)   (b)

FIGURE 6.9

Atomic positions and displacement differences (shown by arrows) for alternative core structures of a screw dislocation with Burgers vector $\frac{1}{2}$[111]. The cores are energetically degenerate and were obtained by computer simulation using an interatomic potential for molybdenum. *(From Vitek and Paidar (2009), Dislocations in Solids, vol. 14, p. 439 (ed. J. P. Hirth), North-Holland. Copyright (2009) with permission from Elsevier.)*

satisfying this symmetry. Examples of these are shown in Figs 6.9 and 6.10. The circles in these plots represent atom positions projected on the (111) plane of the paper. Shading denotes atoms on three adjacent (111) planes: the interplanar spacing is $b/3$. The major atomic displacements are parallel to the dislocation line [111], and the atom projections are essentially the same as for a perfect crystal. In order to represent the [111] *disregistry* of atoms in the core, arrows are drawn between pairs of neighboring atoms on the projections. The length of an arrow is proportional to the difference of the [111] displacements of the two atoms, and scaled such that for a displacement difference of $b/3$, the arrow just runs from one atom to the other. When the difference falls between $b/2$ and b, it is reduced by b. For the isotropic elastic solution (equation (4.11)), the length of the arrows would decrease in inverse proportion to the distance from the core center and would exhibit complete radial symmetry. In the atomic model, the displacements are concentrated on the three {110} planes of the [111] zone. The cores are invariant with respect to the [111] three-fold axis.

The core of Fig. 6.10 is also invariant with respect to the [10$\bar{1}$] two-fold axis (Fig. 6.8). The one in Fig. 6.9(a) is not, so that an equivalent core obtained by applying a rotation of 180° about [10$\bar{1}$] must exist with the same energy: this is the core shown in Fig. 6.9(b). The core structure existing in two symmetry-related forms in Fig. 6.9 is said to be *degenerate* or *polarized*; the core in Fig. 6.10 is *non-degenerate*.

An interpretation of the concentration of atomic disregistry in the degenerate core is that each of the three intersecting {110} planes contains a fault produced by displacement $\frac{b}{3} = \frac{1}{6}$[111]. Similarly, the non-degenerate core can be considered to spread as six unstable faults with displacement $\frac{b}{6}$ on the three {110} planes. Although the core disregistry is distributed as either three $\frac{b}{3}$ or six $\frac{b}{6}$ cores in this simple picture, these *fractional* dislocations do not bound stable stacking faults. Thus, the core structure predicted by computer simulation depends rather sensitively on the shape of the γ-surface, and different interatomic potentials for the same metal have been found to produce both core types, as is apparent from Figs 6.9 and 6.10. Calculations using *ab initio* methods (section 2.4) are now considered more reliable and indicate that the screw dislocation has the non-degenerate form in iron, molybdenum, tantalum and tungsten.

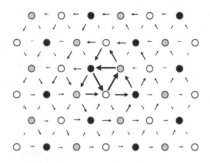

FIGURE 6.10

Atomic positions and displacement differences (shown by arrows) for the non-degenerate core structure of a screw dislocation with Burgers vector $\frac{1}{2}$[111]. The structure was obtained by computer simulation using an interatomic potential for molybdenum. *(From Vitek and Paidar (2009), Dislocations in Solids, vol. 14, p. 439 (ed. J. P. Hirth), North-Holland. Copyright (2009) with permission from Elsevier.)*

Glide of both types of screw core has been studied by applying stress to model crystals. The effect of resolved shear stress is to remove the symmetry associated with the [10$\bar{1}$] two-fold axis from the non-degenerate core (Fig. 6.10). Under an increasing [111](10$\bar{1}$) shear stress tending to move the dislocation to the right, the core disregistry extends to the right in a similar form to the $\frac{b}{3}$ fractional dislocation on the right in Fig. 6.9(a). Under the reverse stress, the core extends to the left, as on the left-hand side of the degenerate core in Fig, 6.9(b). Thus, the behavior under stress is similar for the degenerate and non-degenerate screw core structures. Although details of the core changes are dependent on the interatomic potentials, it is found that under pure shear stress slip occurs on the {110} planes with the asymmetry described above. Furthermore, computer simulation shows that the screw core responds differently to stresses with different non-shear components, in good agreement with the effects of compression and tension found in experiment.

Similar studies of non-screw dislocations, on the other hand, show they have cores which are planar in form on either {110} or {112} but do not contain stable stacking faults. Like their face-centered cubic counterparts, they are not sensitive to the application of non-shear stresses, and they glide at much lower shear stresses than the screw dislocation.

Glide of the screw dislocation therefore controls yielding in the body-centered cubic metals. Glide of a straight screw dislocation requires a high applied stress, and so it moves by a kink-pair mechanism (section 10.3). The effect of this is apparent in the strong temperature-dependence of the yield stress in these metals. For example, the critical resolved shear stress (CRSS) for single crystals of iron falls from approximately 250 to 20 MNm^{-2} between 77 and 295 K.

Another set of perfect dislocations in the body-centered cubic metals are those with Burgers vector $\langle 001 \rangle$, which has length a compared with $\sqrt{3}a/2$, where a is the lattice parameter. $\langle 001 \rangle$ dislocations are occasionally observed in dislocation networks and are believed to occur from the reaction of two perfect $\frac{1}{2}\langle 111 \rangle$ dislocations:

$$\frac{1}{2}\left[1\bar{1}1\right] + \frac{1}{2}\left[11\bar{1}\right] \rightarrow [100] \qquad (6.8)$$

Computer modeling of the edge dislocation has shown that the CRSS for glide is much larger than that for the $\frac{1}{2}\langle 111 \rangle$ edge dislocation. $\langle 001 \rangle$ slip is therefore unlikely to take part in plastic deformation at room temperature and below. However, calculations for iron using anisotropic elasticity theory, rather than the more approximate isotropic elasticity theory used in Chapter 4, show that the elastic energy of a $\langle 001 \rangle$ dislocation falls strongly as temperature increases towards 912°C, the $\alpha-\gamma$ transition temperature at which the stable crystal structure of iron changes from body-centered cubic to face-centered cubic. It is possible, therefore, that dislocations with $b = \langle 001 \rangle$ play a significant role in the high-temperature plasticity of α-iron.

Deformation twinning is observed in all the body-centered cubic transition metals when they are deformed at low temperature and/or high strain rate. As noted with respect to the hexagonal close-packed metals in section 6.2, and described in more detail in section 9.7, the atomic displacements that allow growth of a twin occur by the glide of steps (twinning dislocations) over the twin boundary plane. For the body-centered cubic metals twinning occurs on $\{112\}\langle 111 \rangle$ systems. In section 1.2 it was shown that the stacking sequence of $\{112\}$ planes in the body-centered cubic structure is $ABCDEFAB\ldots$ The homogeneous shear required to produce a twin is $1/\sqrt{2}$ in a $\langle 111 \rangle$ direction on a $\{112\}$ plane. This shear can be produced by a displacement of $\frac{1}{6}\langle 111 \rangle$ on every successive $\{112\}$ plane i.e. the Burgers vector of a twinning dislocation has the form $\frac{1}{6}\langle 111 \rangle$ and *the step height is* $a/\sqrt{6}$. It is seen from Fig. 1.6(b) that if all the atoms in, say, an E layer and above are translated by $\frac{1}{6}\left[\bar{1}11\right]$, then E shifts to C, F shifts to D, etc., and the new sequence is $ABCDCDE\ldots$ A second translation on the adjacent plane displaces D to B, E to C, etc., resulting in $ABCDCBC\ldots$ Repetition of this translation on successive planes gives $ABCDCBA\ldots$, which is the stacking of a twinned crystal. (This process is described in more detail in section 9.7.)

From the atom positions in Fig. 1.6(b), it is also clear that the single translation $\frac{1}{6}\left[\bar{1}11\right]$ which displaces E to C, F to D, etc., produces a different result from the translation $\frac{1}{6}\left[1\bar{1}\bar{1}\right]$ of opposite sense. The latter moves E to D and produces an untwinned structure of high energy. There is thus an asymmetry with respect to $\frac{1}{6}\langle 111 \rangle$ translations in the *twinning* and *anti-twinning sense* on $\{112\}$ planes.

Dislocation loops formed by interstitials and vacancies are an important product of radiation damage. There are no close-packed planes in the

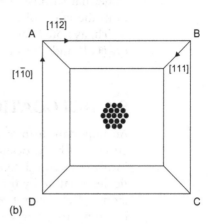

FIGURE 6.11

(a) Computer simulation of a cluster of 19 self-interstitial atoms in iron at 260 K showing the position of the cluster at two different times (1 ps = 10^{-12} seconds). The light and dark spheres are vacant sites and interstitial atoms respectively. The cluster consists of [111] crowdions and moves along its [111] axis by forward and backward jumps of the individual crowdions, which remain bound together during this one-dimensional migration. (b) Cluster in projection along the migration axis. *(Courtesy Yu. N. Osetsky.)*

body-centered cubic structure, and it has been suggested that the loops nucleate with Burgers vector $\frac{1}{2}\langle 110 \rangle$ on the $\{110\}$ planes, which are the most densely packed. In the absence of stable stacking faults on these planes, the partial dislocation loops would shear at an early stage of growth to become perfect by one of two reactions:

$$\frac{1}{2}\langle 110 \rangle + \frac{1}{2}\langle 001 \rangle \rightarrow \frac{1}{2}\langle 111 \rangle \tag{6.9}$$

$$\frac{1}{2}\langle 110 \rangle + \frac{1}{2}\langle 1\bar{1}0 \rangle \rightarrow \langle 100 \rangle \tag{6.10}$$

Since dislocation energy is proportional to b^2, the resultant dislocation in reaction (6.9) should be favored and loops with this Burgers vector have been observed in many metals. However, atomic-scale computer simulation (section 2.4) of damage production in displacement cascades shows that small clusters of vacancies and self-interstitial atoms nucleate directly as loops with Burgers vector $\frac{1}{2}\langle 111 \rangle$. An interstitial cluster consists of $\langle 111 \rangle$ *crowdion* interstitials, i.e. each interstitial is an extra atom inserted in a close-packed $\langle 111 \rangle$ atomic row, as illustrated by the computer-generated image of a cluster of 19 interstitials in Fig. 6.11. The distortion is focused along the crowdion axis and simulation predicts that these small loops move easily along their glide prism. The one in Fig. 6.11 has moved by about $7a$.

In α-iron and its alloys an increasing proportion of loops are found to be of $\langle 100 \rangle$ type when the irradiation temperature is raised above 300°C. This is somewhat surprising in view of the magnitude of b^2. It is believed to arise from the effect of temperature on the anisotropic elastic constants of iron, which, as noted above for straight dislocations, modifies the relative stability of $\frac{1}{2}\langle 111 \rangle$ and $\langle 100 \rangle$ dislocations.

6.4 DISLOCATIONS IN IONIC CRYSTALS

An important feature of dislocations in ionic solids is that electrical charge effects can be associated with them. For example, compressive deformation increases electrical conductivity and electric current is produced during plastic deformation by the motion of charged dislocations. Ionic crystals contain atoms of elements from different sides of the periodic table which transfer electrons from one species to the other, producing sets of *cations* and *anions*. One of the simplest forms is the *rocksalt* (*NaCl*) structure shown in Fig. 1.12, in which each anion is surrounded by six cations and vice versa. It has a face-centered cubic Bravais lattice with an anion−cation pair for each lattice point, one ion at 0, 0, 0 and the other at $\frac{1}{2}$, 0, 0. MgO, LiF and AgCl also have this structure. It has been widely studied and its description provides a basis for more complicated systems.

The shortest lattice vector is $\frac{1}{2}\langle 110 \rangle$, and this is the Burgers vector of the dislocations responsible for slip. The principal slip planes are $\{110\}$. Slip steps are also observed on $\{100\}$ and (occasionally) $\{111\}$ and $\{112\}$ planes after high stresses, particularly at high temperatures and in crystals of high polarizability, where the ionic nature of bonding decreases. Cross slip of screw dislocations can occur only by glide on planes other than $\{110\}$, for only one $\langle 110 \rangle$ direction lies in a given $\{110\}$ plane.

Figure 6.12 shows a pure *edge* dislocation with a $\frac{1}{2}[110]$ Burgers vector and $(1\bar{1}0)$ slip plane emerging on the (001) surface. The extra half-plane actually consists of two supplementary half-planes, as shown. The ions in the planes below the surface alternate between those shown in Figs 6.12(a) and (b). The figure serves to illustrate that there is an *effective charge* associated with the point of emergence of the dislocation on the (001) surface. Intuitively, if it is $-q$ in (a) it must be $+q$ in (b). It is readily shown that $q = e/4$, where e is the electronic charge, as follows. In any cube of NaCl in which the corner ions are of the same type, as in Fig. 1.12, the ions of that sign exceed in number those of opposite sign by one. This excess charge of $\pm e$ may be considered as an effective charge $\pm e/8$ associated with each of the eight corners. In a cube with equal numbers of anions and cations, the positive and negative corners neutralize each other. Consider the dislocation of Fig. 6.12(a) to

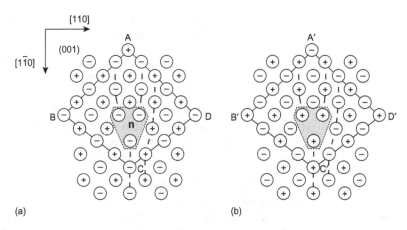

FIGURE 6.12

Edge dislocation in the sodium chloride structure, with the Na$^+$ cations represented by + and the Cl$^-$ anions by −. (a) Initial configuration of surface ions. (b) Configuration after removal of one surface layer. *(From Amelinckx (1979), Dislocations in Solids, vol. II, p. 66, North-Holland.)*

be in a block of crystal *ABCD* bounded by {100} faces. The effective charge of the four corners *A*, *B*, *C*, *D* is $(+e/8 − 3e/8) = −e/4$, and so the net effective charge at the (001) surface is $(−e/4 − q)$. Remove a single layer of ions to expose the new face *A′B′C′D′* (Fig. 6.12(b)). The net effective charge is now $(−e/8 + 3e/8 + q)$, which has been achieved by removing sixteen anions and fifteen cations, i.e. $−e$. Since the initial charge must equal the final charge plus the charge removed,

$$-\frac{e}{4} - q = \frac{e}{4} + q - e \qquad (6.11)$$

and $q = e/4$. The same result holds for emergence on {110} planes.

When the edge dislocation glides on its (1$\bar{1}$0) slip plane, there is no displacement (and hence transport of charge) along the line, and the effective charge of the emergent point does not change sign. For the same reason, *kinks* in edge dislocations bear no effective charge. If the dislocation climbs by, say, removal of the anion labeled *n* at the bottom of the extra half-plane in Fig. 6.12(a), the configuration changes to the mirror image of that in Fig. 6.12(b), and so the effective charge at the emergent point changes sign. It follows that *jogs* carry effective charge, as demonstrated by the illustration in Fig. 6.13 of two *elementary* (or *unit*) jogs of one atom height. The bottom row of ions has an excess charge $−e$, which is effectively carried by the two jogs. Thus, depending on the sign of the end ion of the incomplete row, an elementary jog has a charge $±e/2$. It cannot be neutralized by point defects of integer charge. Charged jogs attract or repel each other electrostatically, and only jogs of an even number times the elementary height are neutral.

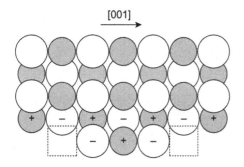

FIGURE 6.13

Extra half-planes of the edge dislocation in Fig. 6.12 with a jog at each end denoted by the squares. *(From Amelinckx, Supplement Vol. 7, Series X, Nuovo Cimento, p. 569, 1958.)*

(Note that in divalent crystals such as MgO, the effective charges are twice those discussed here.)

The formation energies of anion and cation vacancies are in general different, and this results in a higher probability of a jog being adjacent to a vacant site of the lower energy. Dislocations thus have an effective charge per unit length in thermal equilibrium, although this is neutralized in the crystal overall by an excess concentration of vacancies of opposite sign.

The situation with the $\frac{1}{2}\langle 110 \rangle$ *screw* dislocation is more complicated. The ions in any particular $\langle 110 \rangle$ row are of the same sign (see Fig. 1.12), and since displacements are parallel to the Burgers vector, motion of the screw results in displacement of charge parallel to the line. Consequently, both kinks and jogs on screw dislocations are charged, the effective charge being $\pm e/4$ in each case. The effective charge of the point of emergence on {110} and {100} surfaces is $\pm e/8$.

The reason underlying the choice of {110} as the principal slip plane is unclear. It has long been considered that the glide system is determined by the strength of the electrostatic interactions within the dislocation core. This is partly supported by computer calculations. Although stable stacking faults do not exist, the core may spread on the {110} planes to a width of about $6b$, and thus consist of two fractional dislocations of Burgers vector $\frac{1}{4}\langle 110 \rangle$ bounding an unstable fault. Spreading on the {100} planes does not occur and is small on {111}. It has also been suggested that, since the ions of the row at the bottom of the extra half-plane of the $\frac{1}{2}\langle 110 \rangle$ edge dislocation all have the same sign for {100} and {111} slip, but alternate in sign for {110} slip, interaction between edge dislocations and charged impurities may be an important factor.

6.5 DISLOCATIONS IN SUPERLATTICES

In many substitutional solid solutions of one element, A, in another, B, the different species of atoms are arranged at random on the atomic positions of the lattice. At a composition, A_xB_{1-x}, for example, any given lattice point is occupied indifferently by either A or B atoms. There are some solid solutions, however, particularly near stoichiometric compositions such as AB, AB_2, AB_3, etc., in which a specific distribution of the atom species can be induced. Below a critical temperature, atoms of one kind segregate more or less completely on one set of lattice positions, leaving atoms of the other kind to the remaining positions. The resulting arrangement can be described as a lattice of A atoms interpenetrating a lattice of B atoms. A random solid solution

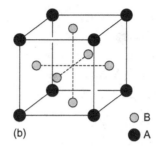

B
A

FIGURE 6.14
Unit cells of the *B2* and *L1$_2$* superlattices.

is changed to an *ordered alloy* with a *superlattice*. Many ordered structures exist. Two possible cubic superlattices produced by alloys of composition *AB* (e.g., CuZn, NiAl) and *AB$_3$* (e.g., Cu$_3$Au, Ni$_3$Al) are shown in Figs 6.14(a) and (b). These structures are given the crystallographic identification *B2* and *L1$_2$*, respectively. In the disordered state, CuZn (β-brass) is body-centered cubic and Cu$_3$Au is face-centered cubic. In the ordered form, both are based on a simple cubic Bravais lattice, one with two atoms per unit cell, the other with four. During the nucleation and growth of ordered domains in a disordered crystal, the lattice parameter change is usually sufficiently small for the atomic planes to remain continuous. Thus, when domains meet, the *A* and *B* sublattices are either in phase, i.e. in 'step', with each other or out of phase. The latter condition results in an *antiphase boundary* (APB). It has a characteristic energy because the nearest-neighbor coordination of the superlattice is destroyed: typical energies are similar to those of stacking faults, i.e. $\sim 10-100$ mJ m^{-2}.

Antiphase boundaries also arise in the core of dislocations in ordered alloys and this is of considerable technological significance because of their influence on the high temperature mechanical properties of these materials. At low temperature, for example, *L1$_2$* alloys behave like the face-centered cubic metals in that the critical resolved shear stress (CRSS) is almost independent of temperature, but as the temperature is increased the CRSS actually increases, an effect known as the *yield stress anomaly*. If the order–disorder transition temperature for the alloy is high enough, the yield stress reaches a peak. This occurs between 800 and 1000 K in the case of Ni$_3$Al and allows Ni-based superalloys that contain Ni$_3$Al precipitates (γ' phase) in a disordered matrix (γ phase) to be used in high temperature applications, e.g., turbine blades. The key to this property is the core structure of the dislocations responsible for slip.

In disordered *L1$_2$* alloys, dislocation behavior is similar to that described for face-centered cubic metals in Chapter 5, e.g., the unit slip vector is $\frac{1}{2}\langle 110\rangle$. In the ordered state (Fig. 6.14(b)), $\frac{1}{2}\langle 110\rangle$ vectors are not lattice translation

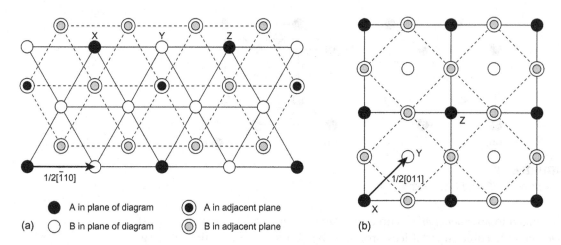

FIGURE 6.15

Arrangement of atoms in two adjacent atomic planes in an $L1_2$ superlattice: (a) (111) planes and (b) (100) planes.

vectors and so gliding dislocations leave behind a surface of disorder (*APB*). This is illustrated schematically for the $L1_2$ superlattice in Fig. 6.15(a), which shows the atomic arrangement in two adjacent {111} planes. Displacement of one layer by $\frac{1}{2}\langle110\rangle$ with respect to the other shifts X to Y and creates an *APB* in which A atoms occupy nearest-neighbor sites to each other. Order may be restored, however, by a second displacement $\frac{1}{2}\langle110\rangle$ which takes the atom originally at X to Z.

Unlike the face-centered cubic metals, the $L1_2$ alloys are also commonly observed to slip on the {100} planes. In this case, see Fig. 6.15(b), the shift X to Y leaves nearest-neighbor bonds across the *APB* unchanged and the second-neighbor changes are the major contribution to the *APB* energy, implying that the *APB* energy is lower on {100} than {111}. Again, the second shift Y to Z restores order.

Thus, the perfect dislocation moving on either {111} or {100} planes of the $L1_2$ superlattice consists of two $\frac{1}{2}\langle110\rangle$ *superpartial* dislocations joined by an *APB*. This *superdislocation* has a Burgers vector $\langle110\rangle$, which is a lattice translation vector. The superdislocation is similar to two Shockley partials joined by a stacking fault (section 5.3), for the spacing in equilibrium is given by a balance between the elastic repulsive force between the two superpartials and the opposing force due to the *APB* energy. Furthermore, if the two dislocations are each dissociated into $\frac{1}{6}\langle112\rangle$ Shockley partials in the disordered phase, as is possible in the face-centered cubic structure, they may retain this form in the ordered $L1_2$ lattice. The superdislocation would then consist of two extended dislocations connected by an *APB*. The slip system of the

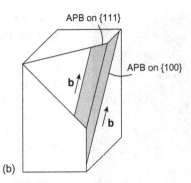

FIGURE 6.16

(a) Screw superpartial pair with an *APB* on a {111} plane. (b) Cross slip onto a {100} plane, starting the formation of a Kear–Wilsdorf lock.

ordered structure is apparently stabilized by the dislocation behavior in the disordered state, for the Burgers vector of the superdislocation is not the shortest lattice vector, which is $\langle 100 \rangle$ in both structures in Fig. 6.14. In the alloy of Fig. 6.14(a), which is body-centered cubic in the disordered state, the Burgers vector is $\langle 111 \rangle$, given by two $\frac{1}{2}\langle 111 \rangle$ components. The requirement that dislocations in superlattices travel in pairs separated by an *APB* provides the strengthening mechanism referred to above. It arises because of a 'locking' mechanism that affects the superpartial pair in the screw orientation.

It was seen in section 5.3 that when the perfect $\frac{1}{2}\langle 110 \rangle$ dislocation in a face-centered cubic metal dissociates into two Shockley partials, it is restricted to glide in one {111} plane because the Burgers vector $\frac{1}{6}\langle 112 \rangle$ of a partial lies in only one {111} plane. The glide plane can only change when constriction occurs. The same behavior will apply to each superpartial in an $L1_2$ alloy if there is a tendency to split into Shockley partials. The two $\frac{1}{2}\langle 110 \rangle$ dislocations will be separated by a {111} *APB*, as shown for a screw superdislocation in Fig. 6.16(a). If one of the superpartials is not dissociated, it can glide on a {100} plane, as illustrated in Fig. 6.16(b). This cross slip from the {111} ('octahedral') plane to the {100} ('cube') plane may be energetically favorable if the *APB* energy is lower on {100} than {111}, which is the implication of the simple geometrical picture of nearest-neighbor coordination described above. Furthermore, depending on the elastic anisotropy of the crystal, the force (or torque) one partial exerts on the other as a result of its stress field can enhance the stability of this arrangement.

When cross slip of a screw superdislocation onto {100} planes occurs, glide of the remainder of the dislocation on the {111} system is restricted because the Peierls stress resisting glide on {100} is much higher than that on {111}.

FIGURE 6.17

Dislocation structure in Ni$_3$(Al,Hf) deformed at 400°C. The (111) slip plane is the plane of the paper and the [10$\bar{1}$] slip direction is indicated by **b**. The superdislocations are elongated in the screw direction and are bowed out in the (010) cube cross-slip plane. The edge segments are shorter. The superdislocations consist of two $\frac{1}{2}$[10$\bar{1}$] superpartials with a spacing of a few nm. *(Courtesy P. Veyssiere.)*

Thus, the applied stress has to be increased to maintain plastic flow. The {100} cross-slipped segment is known as a *Kear–Wilsdorf lock*. The effect of temperature that results in the yield stress anomaly referred to above probably arises from the ease with which a $\frac{1}{2}\langle 110 \rangle$ superpartial can cross slip onto a {100} plane. If it is dissociated into two $\frac{1}{6}\langle 112 \rangle$ Shockley partials on a {111} plane, it will have to constrict, as mentioned above, and the energy barrier for that can be overcome by thermal activation. The result is increased hardening due to cross slip with increasing temperature, as seen experimentally. An example of the dislocation structure in a crystal of Ni$_3$(Al,Hf) deformed at 400°C by slip on a single slip system is shown in the transmission electron microscope image taken under dark-field conditions (section 2.2) in Fig. 6.17. Not all $L1_2$ alloys exhibit anomalous hardening, however, and this probably results from a higher ratio of the *APB* energy on {100} to that on {111} and a lower value of the elastic torque. Finally, it should be noted that sessile screw configurations equivalent to the Kear–Wilsdorf lock have been found in ordered alloys with other crystal structures.

6.6 DISLOCATIONS IN COVALENT CRYSTALS

The covalent bond formed by two atoms sharing electrons is strongly localized and directional, and this feature is important in determining the characteristics of dislocations. Of the many covalent crystals, the cubic structure of diamond, silicon and germanium is one of the simplest and most widely studied. Compounds such as gallium arsenide (GaAs) have the same atomic arrangement. Dislocations in these semiconductors affect both mechanical and electrical properties.

The space lattice is face-centered cubic with two atoms per lattice site, one at 0, 0, 0 and the other at $\frac{1}{4}, \frac{1}{4}, \frac{1}{4}$ (Fig. 6.18). In GaAs one atom would be gallium and the four nearest-neighbors arsenic. Each atom is tetrahedrally bonded to four nearest-neighbors, and the shortest lattice vector $\frac{1}{2}\langle 110 \rangle$ links a second-neighbor pair. The close-packed {111} planes have a six-fold stacking sequence *AaBbCcAaBb*... as shown in Fig. 6.19. Atoms of adjacent layers of the same letter such as *Aa* lie directly over each other, and planar stacking

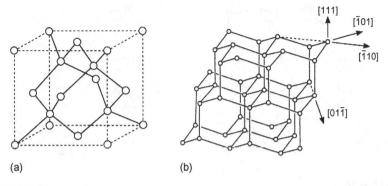

FIGURE 6.18

(a) Diamond-cubic unit cell. (b) Illustration of the structure showing the tetrahedral bonds and important crystallographic directions. The (111) planes are horizontal.

FIGURE 6.19

$(01\bar{1})$ projection of the diamond-cubic lattice showing the stacking sequence of the (111) planes and the shuffle and glide planes defined in the text.

faults arising from insertion or removal of such pairs do not change the tetrahedral bonding. By reference to the face-centered cubic metals, the intrinsic fault has stacking sequence *AaBbAaBbCc...* and the extrinsic fault has *AaBbAaCc AaBb...* Faults formed between adjacent layers of the same letter do not restore tetrahedral bonding and have high energy.

Perfect dislocations have Burgers vector $\frac{1}{2}\langle 110 \rangle$ and slip on $\{111\}$ planes. They usually lie along $\langle 110 \rangle$ directions at $0°$ or $60°$ to the Burgers vector as a result of low core energy in those orientations (see Fig. 8.6). From consideration of dislocations formed by the cutting operations of section 3.2, two

(a)   b   (b)   b

FIGURE 6.20
Perfect 60° dislocations of (a) the glide set and (b) the shuffle set. The extra half-planes are in the
lower part of each drawing. *(After Hirth and Lothe (1968),* Theory of Dislocations, *McGraw-Hill.)*

dislocation types may be distinguished. The cut may be made between layers
of either different letters, e.g., *aB*, or the same letter, e.g., *bB*. Following Hirth
and Lothe the dislocations produced belong to either the *glide set* or the *shuffle set*, as denoted in Fig. 6.19. Diagrammatic illustrations of the two sets are
shown in Fig. 6.20. The *dangling bonds* formed by the unsaturated free bond
per atom along the core are apparent.

Dislocations of both sets are glissile and can dissociate, but the mechanism
of dissociation is different for the two cases. In the glide set, the perfect dislocation dissociates into two $\frac{1}{6}\langle 112 \rangle$ Shockley partials separated by the intrinsic stacking fault, as in the face-centered cubic metals. Dissociation of the
shuffle dislocation is not so simple because of the absence of low-energy
shuffle faults. It occurs by the nucleation and glide of a Shockley partial of
the glide type between an adjacent pair of $\{111\}$ layers. This results in a
fault of the glide set bounded on one side by a Shockley partial and on the
other by a Shockley partial and, depending on whether the glide fault is
above or below the shuffle plane, a row of interstitials or vacancies. It is
probable that this dislocation is less mobile than the glide-set dislocation
because movement of the row of point defects within the core can only
occur during slip by shuffling: hence the nomenclature. Climb, which
involves point defect absorption or emission, transforms shuffle-set dislocations to glide-set dislocations, and vice versa. The stacking-fault energy in silicon and germanium is sufficiently low ($\simeq 50$ mJm^{-2}) for dissociation to be
resolved directly in the transmission electron microscope (Fig. 2.5).

Computer calculations of the core energy of dislocations in model silicon
crystals suggest that the energy is reduced by *bond reconstruction*, a process
in which dangling bonds reform with others so that all atoms retain

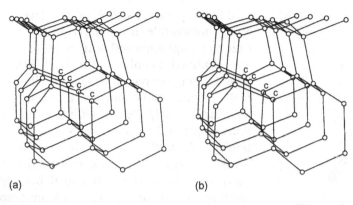

(a)   (b)

FIGURE 6.21

Two possible core configurations for the 30° partial of the glide set in silicon. (a) Dangling bonds occur at atoms $CCCC$. (b) Reconstructed core. *(From Marklund, Phys. Stat. Sol. (b), 92, 83, 1979.)*

approximately tetrahedral coordination. This is shown schematically in Fig. 6.21 for the glide-set partial with $\mathbf{b} = \frac{1}{6}\langle 11\bar{2}\rangle$ at 30° to the line direction. Reconstruction occurs by the dangling bonds at $CCCC$ rebonding in pairs along the core. The structure within the core affects the electron energy levels and, therefore, the electrical properties of crystals containing dislocations. Electronic paramagnetic measurements on deformed silicon indicate that most sites along dislocation cores are reconstructed.

The Peierls stress for glide of straight dislocations in these crystals at low temperature is high and so motion occurs by kink-pair nucleation and migration (see section 10.3). The situation is more complex than in metals because bond reconstruction in the core of dislocations of different orientation results in the occurrence of kinks of a variety of types. Computer simulation has been used to investigate their form and energy, and to assist speculation about the relative importance of perfect and partial dislocations under different conditions of stress and temperature (see Further Reading).

6.7 DISLOCATIONS IN LAYER STRUCTURES

There is a large group of materials with a pronounced layer-type structure which can arise in two ways. Firstly, when the binding forces between atoms in the layers are much stronger than the binding forces between atoms in adjacent layers, as, for example, in graphite. Secondly, when the arrangement of the atoms in complex molecular structures results in the formation of two-dimensional sheets of molecules as, for example, in talc and mica. There are a number of important consequences of the *layer structure*. Slip occurs

1μm

FIGURE 6.22

Electron transmission micrograph of dislocation ribbons in talc. (*From Amelinckx and Delavignette (1962)*, Direct Observation of Imperfections in Crystals, *p. 295, Interscience.*)

readily in planes parallel to the layers and is almost impossible in non-layer planes. The dislocation arrangements and Burgers vectors are therefore confined mainly to the layer planes. The weak binding between layers can result in stacking faults of low energy and hence unit dislocations are usually widely dissociated into partial dislocations. Figure 6.22 shows an example of dislocations in talc. In this material the unit dislocations dissociate into four component partial dislocations. The dislocations appear as ribbons lying in the layer planes, and in some circumstances the electron diffraction conditions allow all four partials to be observed.

Many studies have been made of crystals with layer structures. They are particularly convenient to study experimentally because uniformly thick specimens for transmission electron microscopy can be obtained simply by cleavage along the layer planes. The geometry of the Burgers vectors and dislocation reactions are described using the methods developed in preceding sections.

FIGURE 6.23

(a) The CH_2 chain structure of the polyethylene molecule. (b) (001) projection of the unit cell of orthorhombic polyethylene.

6.8 DISLOCATIONS IN POLYMER CRYSTALS

Plastic deformation of crystalline and semi-crystalline polymers involves the mechanisms well established for other crystalline solids, namely dislocation glide, deformation twinning and, in some cases, martensitic transformations. As in preceding sections, dislocation behavior can be described after making allowances for the structure. Polymer crystals are distinguished by having strong covalent bonding in the direction of the molecular chain axis, and weak van der Waals bonding in the transverse directions. This is illustrated for the orthorhombic phase of polyethylene in Fig. 6.23, which has been the subject of much investigation. As a result of the relative stiffness of the molecules, the important dislocations lie along the [001] direction. The shortest lattice vector is [001], and the screw dislocation produces chain-axis slip by gliding as a perfect dislocation on the (100),

[001]

[010]

[001]

(a)

(b)

○ carbon   ● hydrogen

(010) and {110} planes. Transverse slip results from the glide of edge dislocations, and although the shortest lattice vectors in the transverse direction are [010] and [100], the slip vector is $\langle 110 \rangle$. This is because the perfect $\langle 110 \rangle$ dislocation can dissociate into two Shockley partials with Burgers vector approximately $\frac{1}{2}\langle 110 \rangle$ bounding a stacking fault on a {110} plane. The fault is believed to have low energy (≤ 10 mJ m^{-2}), but the [001] screw is prevented from dissociating by Frank's rule (section 4.4).

In bulk polyethylene, the (001) surfaces of a crystal consist of the folds formed by the molecules leaving and re-entering the crystal. The folds are aligned along certain crystallographic directions, and this influences the dislocation behavior within the crystals by favoring slip planes containing the folds.

FURTHER READING

Body-centered Cubic and Hexagonal Metals

Bacon, D. J. (Ed.) (1991) Adriatico Research Conference on 'Defects in Hexagonal Close-packed Metals', *Phil Mag A* 63:819–1113.

Christian JW: Some surprising features of the plastic deformation of b.c.c. metals and alloys, *Metall Trans A* 14:1233, 1983.

Domain C, Besson R, Legris A: Atomic-scale *ab initio* study of the Zr–H system: II. Interaction of H with plane defects and mechanical properties, *Acta Mater* 52:1495, 2004.

Dudarev SL, Bullough R, Derlet PM: Effect of the α-γ phase transition on the stability of dislocation loops in bcc iron, *Phys Rev Lett* 100:135503, 2008.

Eyre BL, Loretto MH, Smallman RE: Electron microscopy studies of point defect clusters in metals. In Smallman RE, Harris JE, editors: *Vacancies '76*, London, 1977, The Metals Society, p. 63.

Griffiths M: Evolution of microstructure in h.c.p. metals during irradiation, *J Nucl Mater* 205:225, 1993.

Christian JW, Mahajan S: Deformation twinning, *Prog Mater Sci* 39:1, 1995.

Vitek V, Paidar V: Non-planar dislocation cores: A ubiquitous phenomenon affecting mechanical properties of crystalline materials. In Hirth JP, editor: *Dislocations in solids*, vol 14, 2008, North-Holland, p. 439.

Yasi JA, Nogaret T, Trinkle DR, Qi Y, Hector LG, Curtin WA: Basal and prism dislocation cores in magnesium: comparison of first-principles and embedded-atom-potential methods predictions, *Modell Simul Mater Sci Eng* 17:055012, 2009.

Other Crystal Structures

Amelinckx S: Dislocations in particular structures. In Nabarro FRN, editor: *Dislocations in solids*, vol 2, Amsterdam, 1979, North-Holland, p. 66.

Barber DJ, Wenk H-R, Hirth G, Kohlstedt DL: Dislocations in minerals. In Hirth JP, Kubin L, editors: *Dislocations in solids*, vol 16, 2010, North-Holland, p. 171.

Bonneville J, Caillard D, Guyot P: Dislocations and plasticity of icosahedral quasicrystals. In Hirth JP, editor: *Dislocations in solids*, vol 14, 2008, North-Holland, p. 251.

Delavignette P: Dissociation and plasticity of layer crystals, *J de Physique Colloq* 35(C7):181, 1974.

Haasen P: Dissociation of dislocations and plasticity of ionic crystals, *J de Physique Colloq* 35 (C7):167, 1974.

Mitchell TE, Heuer AH: Dislocations and mechanical properties of ceramics. In Nabarro FRN, Hirth JP, editors: *Dislocations in solids*, vol 12, 2004, North-Holland, p. 339.

Nabarro FRN, Duesbery MS, editors: *'L1$_2$ ordered alloys': vol 10 Dislocation in solids*, 1996, North-Holland.

Rabier J, Pizzagalli L, Demenet JL: Dislocations in silicon at high stress. In Hirth JP, Kubin L, editors: *Dislocations in solids*, vol 16, 2010, North-Holland, p. 47.

Schall P, Spaepen F: Dislocations in colloidal crystals. In Hirth JP, Kubin L, editors: *Dislocations in solids*, vol 15, 2009, North-Holland, p. 233.

Spence JCH: Experimental studies of dislocation core defects. In Nabarro FRN, Hirth JP, editors: *Dislocations in solids*, vol 13, 2007, North-Holland, p. 419.

Sprackling MT: *The plastic deformation of simple ionic crystals*, 1976, Academic Press.

Young RJ: Deformation and fracture of polymer crystals, in *Developments in polymer fracture*, London, 1979, Applied Science Publishers, p. 223.

CHAPTER 7

Jogs and the Intersection of Dislocations

7.1 INTRODUCTION

It has been shown that dislocations glide freely in certain planes under the action of an applied shear stress. Since even well-annealed crystals usually contain a network of dislocations, it follows that every slip plane will be threaded by dislocations and a dislocation moving in the slip plane will have to intersect the dislocations crossing the slip plane. The latter are called *'forest dislocations'*. As plastic deformation proceeds, slip occurs on other slip systems and the slip plane of one system intersects slip planes of the other systems, thus increasing the number of forest dislocations. The ease with which slip occurs depends, to a large degree, on the way the gliding dislocations overcome the barriers provided by the forest dislocations. Since the dislocation density in a crystal increases with increasing strain, the intersection processes affect the rate at which the crystal hardens as it is strained. The elementary features of the intersection process are best understood by considering the geometry of the intersection of straight dislocations moving on orthogonal slip planes.

7.2 INTERSECTION OF DISLOCATIONS

The intersection of two edge dislocations with Burgers vectors at right angles to each other is illustrated in Fig. 7.1. An edge dislocation XY with Burgers vector $\mathbf{b}_1$ is gliding in plane P_{XY}. It cuts through dislocation AB with Burgers vector $\mathbf{b}_2$ lying in plane P_{AB}. Since the atoms on one side of P_{XY} are displaced by $\mathbf{b}_1$ relative to those on the other side when XY passes, the intersection results in a jog PP' (see section 3.6) parallel to $\mathbf{b}_1$ in dislocation AB. The jog is part of the dislocation AB and has a Burgers vector $\mathbf{b}_2$, but the length of the jog is equal to the length of $\mathbf{b}_1$. The Burgers vector of dislocation AB is parallel to XY and no jog is formed in the dislocation XY. The overall length of the dislocation AB is increased by b_1. Since the energy per unit length of a

137

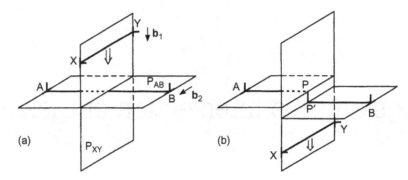

FIGURE 7.1
Intersection of edge dislocations with Burgers vectors at right angles to each other. (a) A dislocation *XY*
moving on its slip plane P_{XY} is about to cut the dislocation *AB* lying in plane P_{AB}. (b) *XY* has cut
through *AB* and produced a jog *PP'* in *AB*.

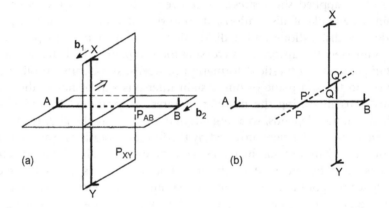

FIGURE 7.2
Intersection of edge dislocations with parallel Burgers vectors. (a) Before intersection. (b) After
intersection.

dislocation is αGb^2 where $\alpha \approx 1$ (equation (4.24)) the energy of the jog is
αGb^3, neglecting the effect of elastic interaction with adjacent dislocation
segments. However, a jog in an undissociated dislocation is a short length of
dislocation with practically no long-range elastic energy and $\alpha \ll 1$. The
energy is largely determined, then, by the core energy of the dislocation (section 4.4). A value of $\alpha = 0.2$ will be used here.

The rule that *when two dislocations intersect each acquires a jog equal in direction
and length to the Burgers vector of the other* may be used to analyze other cases.
The intersection of two orthogonal edge dislocations with parallel Burgers
vectors is illustrated in Fig. 7.2. Jogs are formed on both dislocations. The

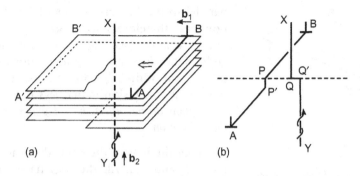

FIGURE 7.3

Intersection of an edge dislocation *AB* with a right-handed screw dislocation *XY*. (a) *AB* moving in its slip plane is about to cut *XY*. Planes threaded by *XY* form a single spiral surface. *AB* glides over this surface and after crossing to *A'B'* the ends do not lie on the same plane. Thus the dislocation must contain a jog *PP'* as shown in (b).

length of the jog QQ' is equal to b_2 and the length of the jog PP' is equal to b_1. The increase in energy as a result of the interaction is twice that for the example above. The intersections of a screw dislocation with an edge dislocation and a screw dislocation are illustrated in Figs 7.3 and 7.4 respectively. The sign of the screw dislocations is represented by the arrows. For the examples given all the screw dislocations are right-handed, according to the definition given in section 1.4. Jogs are produced on all the dislocations after intersection.

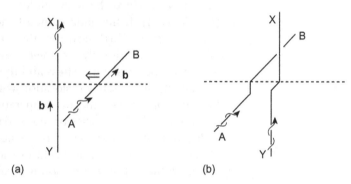

FIGURE 7.4

Intersection of screw dislocations. (a) Before intersection. (b) After intersection.

The length, or 'height', of all the jogs described is equal to the shortest lattice translation vector. These are referred to as *elementary* or *unit jogs*.

7.3 MOVEMENT OF DISLOCATIONS CONTAINING ELEMENTARY JOGS

Consider the jog segment PP' formed on the edge dislocation AB in Fig. 7.1; it is perpendicular to its Burgers vector $\mathbf{b}_2$ and is therefore an *edge dislocation*. The slip plane defined by dislocation AB has a step in it, but the Burgers vector is at all times in the slip plane. Thus, the jog will glide along with the dislocation. The jogs formed on the edge dislocations XY and AB in Fig. 7.2 are

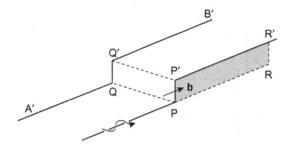

FIGURE 7.5
Movement of a jog on a screw dislocation. Jog *PP′* has a Burgers vector normal to *PP′* and is, therefore, a short length of edge dislocation. The plane defined by *PP′* and its Burgers vector is *PP′RR′* and is the plane in which *PP′* can glide. Movement of the screw dislocation to *A′QQ′B′* would require climb of the jog from *PP′* to *QQ′*.

parallel to the Burgers vectors of the dislocation and therefore have a *screw orientation*. They are *kinks* (see Fig. 3.17) lying in the glide planes P_{XY} and P_{AB} and do not impede the motion of the edge dislocations. Thus, an important conclusion is that *jogs in pure edge dislocations do not affect the subsequent glide of the dislocation.*

Consider the jogs in the screw dislocations in Figs 7.3 and 7.4. All the jogs have an *edge character*. Since an edge dislocation can glide freely only in the plane containing its line and Burgers vector, the only way the jog can move by slip, i.e. *conservatively*, is along the axis of the screw dislocation as illustrated in Fig. 7.5. It offers no resistance to motion of the screw provided the screw glides on the same plane, i.e. the jog is a kink (Fig. 3.17(b)). If the glide plane of the dislocation is different from that of the jog (Fig. 3.17(d)), however, the screw dislocation can move forward and take the jog with it only by a *non-conservative process*. As described in section 3.6, this process requires stress and thermal activation, and consequently the movement of the screw dislocation will be temperature-dependent. At a sufficiently high stress and/or temperature, movement of the jog will leave behind a trail of vacancies or interstitial atoms depending on the sign of the dislocation and the direction the dislocation is moving. A jog which moves in such a direction that it produces vacancies is called a *vacancy jog*, and if it moves in the opposite direction it is called an *interstitial jog*.

A screw dislocation can acquire both vacancy- and interstitial-producing jogs during plastic deformation. A critical applied stress is required for dislocation movement. Consider a screw dislocation with an array of jogs along its length (Fig. 7.6(a)). Under an applied shear stress τ acting in the slip plane in the direction of the Burgers vector **b**, the dislocation *bows out* between the jogs to a radius of curvature R given by equation (4.30) (Figs 7.6(b) and (c)). Two relatively closely-spaced jogs experience a net sideways force from the line tension of the dislocation segments meeting at the jogs (Fig. 7.6(c)), making them glide together and resulting in annihilation (as in Fig. 7.6) or formation of a jog of twice the unit length. The remaining jogs will be of approximately uniform spacing, x. The forward force on a given jog is that due to the applied stress on the two dislocation segments (of average length $x/2$) on each side, and is τbx from equation (4.27). Thus, when a point defect is created and the jog moves forward one atomic spacing b, the work done by the applied load is $\tau b^2 x$. If the formation energy of a point defect at

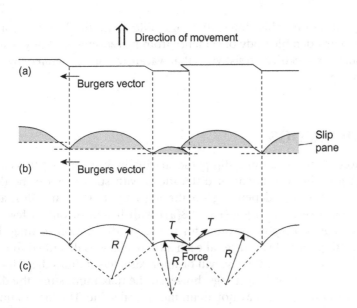

FIGURE 7.6

Movement of a jogged screw dislocation. (a) Dislocation with no applied stress. (b) Dislocation segments pinned by jogs bowing under stress to radius of curvature R. (c) Projection in slip plane showing force on jog due to unbalanced sideways components of line tension T.

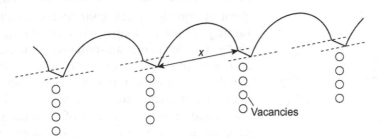

FIGURE 7.7

Glide of a jogged screw dislocation producing trails of point defects.

a jog is E_f, the line moves forward generating point defects (Fig. 7.7) at a critical stress

$$\tau_0 = E_f/b^2 x \qquad (7.1)$$

At temperatures greater than 0 K, thermal activation assists in the formation of point defects and reduces the critical stress of equation (7.1).

Since E_f for interstitials is approximately two to four times that for vacancies in metals (section 1.3), it is unlikely that interstitials are formed by jog dragging. More probably, interstitial jogs glide along the line and combine. There is a considerable body of evidence from measurement of physical properties, such as electrical resistivity, that vacancies are generated by plastic deformation.

7.4 SUPERJOGS

A jog more than one atomic slip plane spacing high is referred to as a *super-* (or *long*) *jog*. The movement of dislocations with such jogs can be divided into three categories, depending on the jog height. They are illustrated in Fig. 7.8. For a very small jog (Fig. 7.8(a)) with height of only a few, say n, atom spacings, it may be possible for the screw dislocation to drag the jog along, creating several vacancies at each atomic plane as described in the preceding section: the critical stress will be approximately n times the value given by equation (7.1). For longer jogs, however, the maximum stress the dislocation line can experience may not attain the critical value. The maximum stress is reached when the segments bowing between jogs of spacing x have their minimum radius of curvature, see equation (4.30), which is $x/2$. If the maximum stress $2\alpha Gb/x$ (given by equation (4.30)) is less than the critical stress nE_f^v/b^2x given by equation (7.1), glide proceeds without vacancy creation. Taking $\alpha = 0.5$ and noting that $Gb^3 \simeq 4$ eV for many face-centered cubic metals and $\simeq 5-10$ eV for many body-centered cubic metals, it is apparent that vacancy generation ceases for n greater than two or three. Under these conditions, the behavior of the dislocation depends on whether or not the superjogs are of intermediate or large height.

At the maximum stress, the bowing segments have radius of curvature $R = x/2$ and are approximately semicircular in shape. The two dislocation arms meeting at a jog are therefore parallel to each other and perpendicular to their Burgers vector. These edge elements of opposite sign adopt a configuration similar to MO and NP in Fig. 7.8(c), and can only pass each other if the force per unit length τb on each due to the applied stress τ exceeds the maximum value of their mutual repulsion force (Fig. 4.14); that is

FIGURE 7.8

Behavior of jogs with different heights on a screw dislocation moving in the direction shown by the double arrow. (a) Small jog is dragged along, creating point defects as it moves. (b) Very large jog — the dislocations *NY* and *XM* move independently. (c) Intermediate jog — the dislocations *NP* and *MO* interact and cannot pass by one another except at a high stress. *(After Gilman and Johnston,* Solid State Physics, *13, 147, 1962.)*

$$\tau b > \frac{0.25\,Gb^2}{2\pi(1-\nu)\gamma} \tag{7.2}$$

where γ is the jog length MN. Jogs of relatively small length will be unable to meet this criterion, and the gliding screw will draw out a *dislocation dipole* consisting of two edge lines of the same Burgers vector and opposite sign, as shown in Fig. 7.8(c). Long jogs, on the other hand, may have sufficiently large values of γ to satisfy condition (7.2). Their two edge arms can effectively behave independently of each other as single-ended sources as illustrated in Fig. 7.8(b) (*see* Chapter 8 on dislocation multiplication). For $\tau = 10^{-3}\,G$, the transition between these two forms of behavior is found from condition (7.2) to occur at $\gamma \simeq 60b$.

Examples of these dislocation arrangements are shown in Fig. 7.9, which is a transmission electron micrograph of a foil from a silicon-iron single crystal

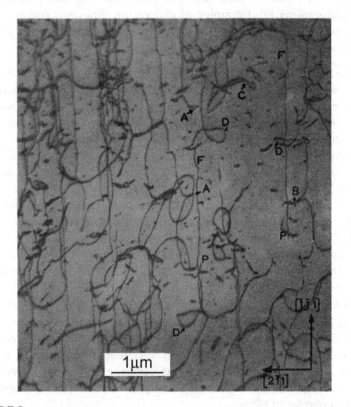

FIGURE 7.9

Transmission micrograph of thin foil of iron 3 per cent silicon alloy parallel to (011) slip plane. *A,* dipole trails. *B* and *C,* pinching off of dipole trails. *D,* single-ended sources at large jogs. *FP,* jogged screw dislocation. *(From Low and Turkalo*, Acta Metall. *10, 215, 1962.)*

FIGURE 7.10
A mechanism for dislocation dipole formation. *(After Tetelman*, Acta Metall. *10, 813, 1962.)*

deformed about 1% ($\tau = 45$ MN m^{-2}, $G = 58$ GN m^{-2}). Slip occurred on only one system, namely $\frac{1}{2}[\bar{1}\bar{1}1](011)$, and the foil is parallel to the (011) slip plane. The $[\bar{1}\bar{1}1]$ and $[2\bar{1}1]$ directions marked on the micrograph are therefore parallel to the screw and edge orientations respectively. The screw dislocations are relatively long and have numerous jogs along their length, e.g. the dislocations *FP*. Examples of dislocation dipoles occur at points *A* and they are aligned approximately in the $[2\bar{1}1]$ edge orientation as expected. Single ended sources have been produced at site *D*.

Double cross slip, which is illustrated in Fig. 3.9, is a ready source of long jogs. The segments of the dislocation which do not lie in the principal slip plane have a predominantly edge character. More generally, any movement of the dislocation out of the slip plane will result in the formation of jogs. Another sequence of events which can lead to superjog and dipole formation during plastic deformation is illustrated in Fig. 7.10. Two dislocations *MM′* and *NN′* with the same Burgers vector **b** but almost opposite line sense are gliding on parallel slip planes of spacing *y* (Fig. 7.10(a)). The force on each due to the stress field of the other acts to reorientate part of their lengths in the glide plane to give parallel edge segments *PP′* and *RR′* of opposite sign (Fig. 7.10(b)). If *y* is sufficiently small, a large shear stress τ is required to separate them (relation (7.2)). Furthermore, if an adjacent part of one line, such as *P′M′*, is close to the screw orientation, it can cross slip to join the other, such as *R′N*. The two segments which join pinch off and leave an edge dipole *PP′R′R* on dislocation *MN′* and superjog *TT′* on dislocation *NM′* (Fig. 7.10(c)). (This reaction is best understood by placing an arrow on each line to denote positive line sense, as in Fig. 7.10(a), and remembering from the Burgers circuit construction (section 1.4) that dislocations of opposite sign have the same Burgers vector but opposite line sense.) Dislocation dipoles are a feature of the early stages of plastic deformation, when slip is confined to one set of planes.

7.5 JOGS AND PRISMATIC LOOPS

Trails of defects and prismatic loops are often produced during plastic deformation. In Fig. 7.9 numerous small loops, many of them elongated, can be seen. This so-called *debris* is left behind by moving dislocations and is a direct result of edge jogs on screw dislocations.

(a)

(b)

(c)

FIGURE 7.11
Formation of dislocation loops from a dislocation dipole. (a) Dislocation dipole. (b) Elongated loop and jogged dislocation. (c) Row of small loops.

Two mechanisms of forming loops are possible. Firstly, by the diffusion and coalescence of vacancies formed at a moving jog, as depicted in Fig. 7.7. If the temperature is sufficiently high to allow diffusion of the defects they can collect together to form a dislocation loop. At low temperatures interstitials diffuse more readily than vacancies and providing interstitials can be formed at an elementary jog, it will be possible to form interstitial dislocation loops. Vacancy loops, however, form only at higher temperatures because of the restricted rate of diffusion. The second mechanism represents a further stage in the development of a dislocation dipole formed either from an intermediate sized jog (Fig. 7.8(c)) or by the interaction of dislocations on parallel slip planes (Fig. 7.10). Two stages in this process are illustrated in Fig. 7.11. The dipole may pinch off from the line by a cross slip mechanism similar to that of dislocations $P'M'$ and $R'N$ in Fig. 7.10(b). The pinching off of dipoles can be seen at points C in Fig. 7.9. Furthermore, the dipole may break up because of the mutual attraction of the positive and negative edge dislocations of its two elongated sides. This results in a row of prismatic loops (Fig. 7.11(c)). It requires climb, but can occur by pipe diffusion at temperatures well below that for volume self-diffusion (section 3.8). This enables material to be re-distributed around the dislocation core. The dislocation loops formed by the breaking up of dipoles can be either vacancy or interstitial loops depending on the sign of the initial jog or dipole.

A prismatic loop is also released when a screw dislocation containing a helical turn breaks free of the turn. The process is the reverse of that discussed in section 3.7, where it is shown that a screw dislocation can absorb a loop and acquire a helical turn (see Fig. 3.23). The dislocation line in the turn can only glide on the glide cylinder of the original loop, i.e. on the cylinder with its axis parallel to the Burgers vector. The glide direction of the screw dislocation is perpendicular to this and so the dislocation will be pinned by the turn. In order for the dislocation to glide away, the screw segments outside the turn will have to bow forward under an applied stress and force the turn to close, thereby reversing the reaction in Fig. 3.23.

Jogs and prismatic loops can also be formed by the movement of dislocations past impenetrable obstacles in dispersion-strengthened alloys. In the *Orowan mechanism* (discussed further in section 10.7), the gliding dislocation wraps around the particle (Fig. 7.12(a)) and by mutual annihilation of segments on the far side leaves behind a *shear loop*, that is, one with its Burgers vector lying in the loop plane. The *Hirsch mechanism* involves cross slip of screw segments in order for the dislocation to bypass the obstacle. Cross slip occurs two or three times, depending on whether or not the line is predominantly edge or screw in character (Figs 7.12(b) and (c)), and one or two prismatic loops are left near the particle. Prismatic loops are observed in transmission electron microscopy of heavily deformed alloys. An Orowan loop produced as in Fig. 7.12(a) can also be induced to cross slip (Fig. 7.12 (d)) and then react with the dislocation to form a prismatic loop and another shear loop. Superjogs are produced in these cross slip processes, but are only permanent features of the line in cases (b) and (d). (Again, a clear understanding of the reactions involved is best obtained by defining the positive line sense, as in the first sketch of each sequence.)

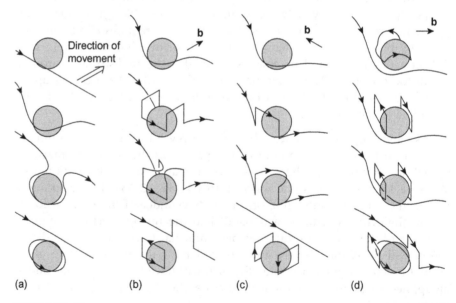

(a)   (b)   (c)   (d)

FIGURE 7.12

Dislocation motion past a particle either (a) without or (b), (c), (d) with cross slip. The Burgers vector in (b), (c), (d) is denoted by **b**. *(After Hirsch and Humphreys (1969), Physics of Strength and Plasticity, p. 189, M. I. T. Press.)*

7.6 INTERSECTIONS OF EXTENDED DISLOCATIONS AND EXTENDED JOGS

The simple geometrical models of intersections illustrated in Figs 7.2–7.5 have to be modified somewhat when the intersecting dislocations are extended into partials, as is expected to be the case in face-centered cubic metals with low stacking fault energy (section 5.3). It is difficult to generalize the problem because of the number of variants of intersections that are possible. One schematic example will be considered to illustrate some of the principles involved. In Fig. 7.13 two dissociated dislocations intersect at right angles, a situation equivalent to Fig. 7.3. The dislocations can only cut through each other by constricting to form perfect unextended dislocations in the region close to the intersection (Fig. 7.13(b)). The dislocations can then separate and elementary jogs are produced on the extended dislocations (Fig. 7.13(c)). It is clear that to form jogs on extended dislocations work will have to be done to constrict the dislocations and so the energy of the jog will depend on the stacking fault energy or, alternatively, on the spacing of the partial dislocations. This problem has been reviewed by Friedel (1964) and Nabarro (1967).

The jogs on the extended dislocations of Fig. 7.13 are shown as single segments for simplicity, but they also can dissociate in crystals where stacking faults are possible. This can have an important effect on dislocation mobility. The configurations are easier to visualize clearly for long jogs, which are able to obey the usual rules for dislocation reactions, and the same forms are often assumed to be valid even for short jogs. The number of possibilities

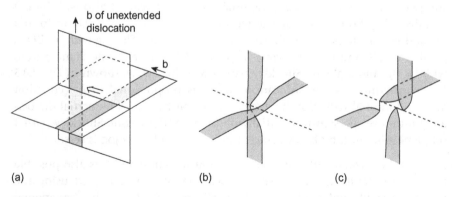

(a)   (b)   (c)

FIGURE 7.13

Intersection of extended dislocations. (a) Extended dislocations moving towards each other. (b) Constriction of dislocations due to the interaction of the elastic strain fields of the leading partial dislocations. (c) Formation of jogs on extended dislocations.

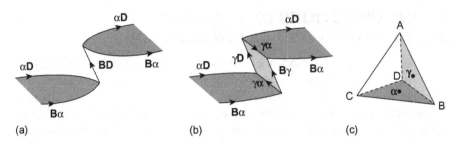

FIGURE 7.14

Formation of a glissile extended jog on an edge dislocation. The jog is constricted in (a) and extended in (b). Positive line senses are denoted by arrows and the Burgers vectors by the letters next to the lines. Orientation of the planes and directions is shown by the Thompson tetrahedron in (c).

is frequently large for a given crystal structure, and two examples for the face-centered cubic case will suffice to indicate the principles involved. The Burgers vectors and planes will be defined with reference to the Thompson tetrahedron described in section 5.4.

Suppose an edge dislocation of Burgers vector ('bv') **BD** extended on plane *BCD* is intersected by a dislocation with bv **BA** to form two acute bends, as illustrated in Fig. 7.14(a). It acquires a jog in the direction *BA* with bv **BD**, which can dissociate on plane *ABD* into two Shockley partials with bv **Bγ** and **γD**, as shown in Fig. 7.14(b). Since Burgers vector is conserved at dislocation junctions (equation (1.9)), this dissociation results in the formation of positive and negative stair-rod partials with bv **γα** along the ⟨110⟩ lines of intersection of the intrinsic stacking faults. The jog will be glissile on the slip plane *ABD*, however. Suppose now a screw dislocation also with bv **BD** and dissociated on *BCD* is intersected by a dislocation with bv **CA** to form a jog and two acute bends, as illustrated in Fig. 7.15(a). The jog with bv **BD** is parallel to *CA* and can dissociate in plane *ACD* into a sessile Frank partial with bv **Bβ** and a glissile Shockley partial with bv **βD**, as shown in Fig. 7.15 (b). Furthermore, since dislocation energy is proportional to b^2 (equation 4.24), additional reduction in energy may occur by the Frank partial dissociating into a sessile stair-rod partial with bv **δβ** and a glissile Shockley partial on plane *ABC* with bv **Bδ** as depicted in Fig. 7.15(c). This jog is sessile.

Jogs may be formed with either acute or obtuse bends, and it is also possible for extrinsic faults to occur. The dissociations can be analyzed using the methods established here, and full details may be found in the original paper of Hirsch (1962) and the reviews of Hirth and Lothe (1982) and Amelinckx (1979).

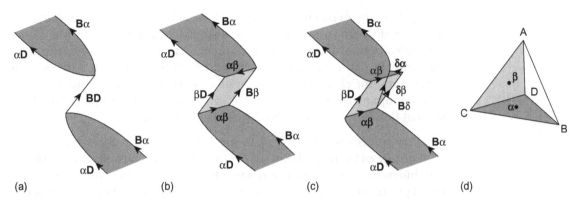

FIGURE 7.15

Formation of a sessile extended jog on a screw dislocation. Jog is constricted in (a), extended in (b), and fully extended in (c). Orientation of the Thompson tetrahedron is shown in (d). Positive line sense and Burgers vector are given for all segments of line.

7.7 ATTRACTIVE AND REPULSIVE JUNCTIONS

The intersection of gliding dislocations with the forest dislocations that intersect their slip planes was treated in section 7.2 as a short-range effect leading to jog and kink formation in the dislocation core. Dislocations can interact at long range, however, by virtue of the stress fields they produce. The treatment of section 4.6 considered this effect for straight, parallel dislocations only, but the same method may be applied to more general orientations. Although conditions for attractive, repulsive and neutral interactions may be determined, the analyses are more complicated because the forces vary with position along each line. Consider the simplest case of two perpendicular screw dislocations A and B parallel to the y- and z-axes, respectively, and having a perpendicular spacing d (Fig. 7.16). The glide force per unit length on A is given by b times the stress σ_{yz} due to B, and the force on B is b times the stress σ_{zy} due to A. These forces are readily obtained from equations (4.13) by suitable choice of axes and origin, and are $Gb^2d/2\pi(d^2+y^2)$ and $Gb^2d/2\pi(d^2+z^2)$ respectively. The force distribution on the dislocations when they have opposite signs is shown schematically

FIGURE 7.16

Interaction forces on two orthogonal screw dislocations A and B: A is left-handed and B right-handed. *(After Hartley and Hirth (1965), Acta Met. 13, 79.)*

in Fig. 7.16: the maximum force on each is $Gb^2/2\pi d$. As a crude approximation, the dislocation shape may be assumed to adopt the profile of the force distribution, but external and internal stresses may modify this substantially. In particular, the line tension which tends to keep a dislocation straight and the change in stress due to change of dislocation shape can be very important for some configurations. However, the analysis based on straight-line interactions elucidates two general points.

First, the forces shown in Fig. 7.16 are repulsive, and work would have to be done, in addition to that required to form a jog, to bring about dislocation intersection. It could be estimated by integrating the forces involved. If the sign of one dislocation is reversed, the forces become attractive and extra work would be required to separate the dislocations after intersection. Thus, dislocation intersections can result in *attractive* and *repulsive* junctions. The extra work is the same for both forms for straight, rigid dislocations. However, real dislocations change shape under their mutual interaction, and whereas the strength of the interaction is thereby reduced for a repulsive junction, it is increased for an attractive one. Attractive junctions are therefore stronger. Second, the two dislocations can react at their junction to form a segment with Burgers vector equal to the sum of the two individual vectors. This may be energetically favorable if the geometry of the vectors is such that b^2 is reduced. The process was illustrated in Section 5.6 for two situations in the face-centered cubic metals in which the product segment is sessile in the interesting glide planes, namely the Hirth and Lomer locks. Two other favorable reactions occur where a sessile lock is not formed. The *glissile reaction* is depicted in Fig. 7.17, where the $\frac{1}{2}\langle 110 \rangle$ Burgers vectors and {111} slip planes are defined by reference to the Thompson tetrahedron (Fig. 7.17(c)). A dislocation of Burgers vector **DC** gliding on plane *BCD* reacts with a forest dislocation with Burgers vector **CB** and lying on plane *ABC*. From the line senses given, it is seen that a segment with Burgers vector **DB** is formed along the line of intersection of the two planes. The segment is glissile on *BCD*, but its ends are restrained by the forest dislocation. The gliding dislocation is therefore held unless the applied stress is sufficient to overcome the reduction in dislocation energy and shrink the segment to zero.

In the case of the *collinear reaction* sketched in Fig. 7.18, the two dislocations MM′ and NN′ in (a) have the same Burgers vector but opposite line sense when pulled into parallel alignment by attraction of their elastic stress fields. The parallel segments annihilate along the line of intersection of the two glide planes, with the result that after the reaction one dislocation runs from M to N′ and the other from N to M′, as in (b). The collinear reaction is considered to be important in work hardening (section 10.8).

FIGURE 7.17

(a) (b) Attractive junction formed by the reaction of a dislocation gliding on plane *BCD* with a forest dislocation on plane ABC. Orientation of the Thompson tetrahedron is shown in (c). (d) Same reaction when the dislocations are extended, showing the formation of an extended node (section 7.8). Burgers vectors in (d) are indicated by the letters against each line.

Reactions that restrict dislocation motion also occur in the body-centered cubic metals: two dislocations with Burgers vector b of the type $\frac{1}{2}\langle 111 \rangle$ can undergo a favorable reaction to form a product segment with $b = \langle 100 \rangle$. For example, with reference to Fig. 7.19(a) and (b), suppose dislocation 1 with $b = \frac{1}{2}[\bar{1}11]$ is gliding on the $(01\bar{1})$ plane and dislocation 2 with $b = \frac{1}{2}[1\bar{1}1]$ is on the $(10\bar{1})$ plane. They can react favorably to form segment AB along the [111] intersection line of the planes: it has $b = [001]$, which does not lie in either of the planes. One or both dislocations will glide away

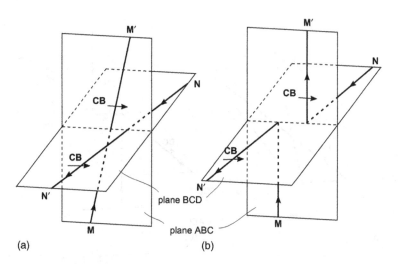

FIGURE 7.18
Attractive interaction resulting in the collinear reaction in which segments of dislocations with the same Burgers vector but opposite line sense annihilate along the line of intersection of their glide planes in a face-centered cubic metal. Orientation of the Thompson tetrahedron is shown in Fig. 7.17(c).

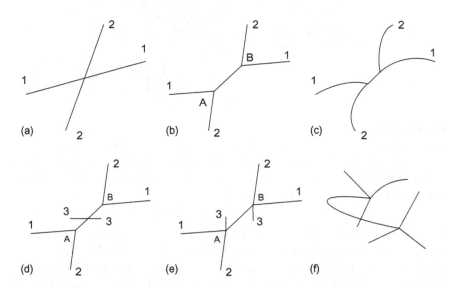

FIGURE 7.19
(a),(b) Formation of $\langle 001 \rangle$ segment AB by reaction of two $\frac{1}{2}\langle 111 \rangle$ dislocations 1 and 2 in a body-centered cubic metal. (c) The segment can unzip under increasing applied stress. (d),(e) A third $\frac{1}{2}\langle 111 \rangle$ dislocation reacts to convert AB to a $\frac{1}{2}\langle 111 \rangle$ screw dislocation. (f) The multi-junction can act as a Frank–Read source if the applied stress is high enough. *(After Bulatov et al., Nature 440, 1174, 2006. Reprinted with permission from Macmillan Publishers Ltd: copyright (2008).)*

when the applied stress is sufficient to cause the junction segment to unzip, as in Fig. 7.19(c). A more complex configuration can arise if another $\frac{1}{2}[111]$ dislocation reacts with AB before it unzips. Dislocation 3 in Fig. 7.19(d) has $b = \frac{1}{2}[11\bar{1}]$ and glides on the $(1\bar{1}0)$ plane, which also contains the $[111]$ direction of AB. Dislocation 3 and AB react according to

$$[001] + \tfrac{1}{2}\left[11\bar{1}\right] = \tfrac{1}{2}[111] \tag{7.3}$$

so that AB becomes a $[111]$ screw segment (Fig. 7.19(e)). This results in a considerable reduction in elastic energy, i.e. three segments each with $b = \frac{1}{2}\langle 111 \rangle$ are reduced to one:

$$\tfrac{1}{2}\left[\bar{1}11\right] + \tfrac{1}{2}\left[1\bar{1}1\right] + \tfrac{1}{2}\left[11\bar{1}\right] = \tfrac{1}{2}[111] \tag{7.4}$$

The nodes at A and B pin the screw segment and so it can act as a Frank–Read source for dislocation multiplication (section 8.5) as the applied stress increases (Fig. 7.19(f)). Dislocation dynamics computer simulation and transmission electron microscopy have shown that the formation of *multi-junctions* of this type is common during plastic deformation and have confirmed that they are strong obstacles to dislocation motion.

7.8 EXTENDED STACKING-FAULT NODES

A particularly interesting dislocation structure, which provides a method for direct experimental measurement of stacking-fault energy γ in face-centered cubic metals, is illustrated in Fig. 7.20. It arises when the dislocations forming an attractive junction such as that shown in Fig. 7.17(b) are extended. The dislocation on plane *BCD* dissociates into two Shockley partials with Burgers vectors (bvs) Dα and αC, and the other on plane *ABC* dissociates into partials with bvs Cδ and δB. The junction segment with bv **DB** dissociates into Shockley partials Dα and αB, as shown in Fig. 7.17(d). The same configuration arises if the two dislocations are considered extended before they intersect, for then partial αC reacts with partial Cδ to form a stair-rod partial $\alpha\delta$ along the line of intersection of the two planes: this in turn combines with partial δB to form the new Shockley partial with Burgers vector αB, which is glissile on plane *BCD* and is caused to bow by surface tension of the stacking fault, as illustrated. Repeated intersections lead to a network of extended and contracted nodes, such as that in a deformed copper–aluminum alloy shown in Fig. 7.21.

The partial dislocation at X in Fig. 7.20 has radius of curvature R and experiences a line tension T tending

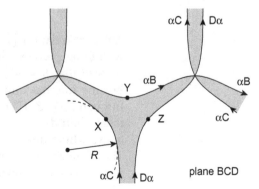

FIGURE 7.20

Extended node in a face-centered-cubic lattice. Shaded area represents a stacking fault. *(After Whelan, Proc. Roy. Soc. A249, 114, 1959.)*

0.5µm

FIGURE 7.21

Transmission electron micrograph of extended nodes in a copper 8 per cent aluminum alloy deformed 5 per cent at room temperature. *(From Swann (1963)*, Electron Microscopy and Strength of Crystals, *p. 131, Interscience.)*

to straighten it. This results in a force T/R per unit length acting towards the center of curvature (equation (4.29)) and is opposed by force γ per unit length due to the 'surface tension' of the stacking fault on the convex side. In equilibrium, therefore,

$$\gamma = \frac{T}{R} \simeq \frac{\alpha G b^2}{R} \qquad (7.5)$$

The equilibrium spacing of Shockley partials well away from the node center is approximately (equation (5.6))

$$d \simeq \frac{G b^2}{4\pi\gamma} \qquad (7.6)$$

Although relation (7.5) takes no account of the effect on R of interactions with other partials and the variation of T with position along the line, it can be seen that $R \approx 6d$. The usual parameter measured experimentally is not R but the internal dimension of the node, e.g. the radius of a circle passing through X, Y, Z, (Fig. 7.20) and this can therefore be an order of magnitude larger than d. Hence, it is possible to measure γ from node observations in alloys where direct measurement of partial spacing is normally impossible. The more-accurate equations employed, and the structure of other forms of nodes, have been reviewed by Amelinckx (1979).

FURTHER READING

Amelinckx S: Dislocations in particular structures. In Nabarro FRN, editor: *Dislocations in solids*, vol 2, Amsterdam, 1979, North-Holland, p. 66.

Bulatov VV, eleven others: Dislocation multi-junctions and strain hardening, *Nature* 440:04658, 2006.

Friedel J: *Dislocations*, 1964, Pergamon.

Hirsch PB: Extended jogs in face-centered cubic metals, *Phil Mag* 7:67, 1962.

Hirsch PB, Humphreys FJ: Plastic deformation of two-phase alloys containing small non-deformable particles. In Argon AS, editor: *Physics of strength and plasticity*, 1969, M. I. T. Press, p. 189.

Hirth JP, Lothe J: *Theory of dislocations*, 1982, Wiley.

Madec R, Devincre B, Kubin L, Hoc T, Rodney D: The role of collinear interaction in dislocation-induced hardening, *Acta Mater* 301:1879, 2003.

Monnet G, Devincre B, Kubin L: Dislocation study of prismatic slip systems and their interactions in hexagonal-close-packed metals: application to zirconium, *Acta Mater* 52:4317, 2004.

Nabarro FRN: *Theory of crystal dislocations*, 1967, Oxford University Press.

Scattergood RO, Bacon DJ: Symmetric stacking fault nodes in anisotropic crystals, *Acta Met* 24:705, 1976.

Seeger A: Jogs in dislocation lines, in *Defects in crystalline solids*, London, 1955, Physical Soc, p. 391.

Veysièrre P: Dislocations and the plasticity of crystals. In Meyers MA, Armstrong RW, Kirchner HOK, editors: *Mechanics and materials: fundamental linkages*, 1999, John Wiley, p. 271.

CHAPTER 8

Origin and Multiplication of Dislocations

8.1 INTRODUCTION

It has been estimated that an upper limit for the contribution the *entropy* of a dislocation makes to the free energy of a crystal is $\sim -2kT/b$ per unit length (Cottrell, 1953). This compares with the strain energy contribution $\sim Gb^2$ (equation (4.24)). Since Gb^3 is typically $\sim$5 to 10 eV and kT at 300 K is 1/40 eV, the net free energy change due to dislocations is positive. Thus, unlike the intrinsic point defects discussed in section 1.3, the thermo-dynamically stable density of dislocations in a stress-free crystal is zero. Nevertheless, apart from crystal whiskers and nanowires, and isolated examples in larger, carefully prepared crystals of materials like silicon, dislocations occur in all crystals. The dislocation density in well-annealed crystals (i.e. crystals which have been heated for a long time close to their melting point to reduce the dislocation density to a low value) is usually about $10^4 \, \text{mm}^{-2}$ ($10^{10} \, \text{m}^{-2}$) and the dislocations are arranged in networks as previously mentioned in Chapter 1 (Fig. 1.22). A similar density of dislocations is present in crystals immediately after they have been grown from the melt or produced by strain anneal techniques. The origin of these dislocations is described in the next section.

When annealed crystals are deformed there is a rapid multiplication of dislocations and a progressive increase in dislocation density with increasing strain. After large amounts of plastic deformation the dislocation density is typically in the range 10^{14} to $10^{15} \, \text{m}^{-2}$. In the early stages of deformation dislocation movement tends to be confined to a single set of parallel slip planes. Later, slip occurs on other slip systems and dislocations moving on different systems interact. The rapid multiplication leads to work hardening (Chapter 10). The main mechanisms of dislocation multiplication are described in this chapter.

157

8.2 DISLOCATIONS IN FRESHLY GROWN CRYSTALS

It is very difficult to grow crystals with low dislocation density because dislocations are readily introduced during the growing process. There are two main sources of dislocations in freshly grown crystals. First, dislocations or other defects may be present in the 'seed' crystals or other surfaces used to initiate the growth of the crystal. Any dislocations in a seed crystal which intersect the surface of the seed on which new growth occurs will extend into the growing crystal. Second, 'accidental' nucleation may occur during the growth process. The main mechanisms which have been suggested are: (a) heterogeneous nucleation of dislocations due to internal stresses generated by impurity particles, thermal contraction, etc.; (b) impingement of different parts of the growing interface; (c) formation and subsequent movement of dislocation loops formed by the collapse of vacancy platelets.

The basic principle involved in the nucleation of dislocations by local internal stresses during growth and subsequent cooling is embodied in the specific example given in section 8.4. High local internal stresses are produced when neighboring parts of the crystal are constrained to change their specific volume. This can occur by neighboring regions expanding or contracting by different amounts due to (a) thermal gradients, (b) change in composition, or (c) change in lattice structure. An additional effect is the adherence of the growing crystal to the sides of the container. When the stress reaches a critical value, about $G/30$, dislocations are nucleated. If this occurs at high temperature the dislocations created will rearrange themselves by climb. It should also be noted that under normal laboratory conditions there is every possibility of isolated vibrations which will affect the growth process and produce additional random stress effects.

The formation of dislocations by impingement is demonstrated during the coalescence of two adjacent dendrites in the growing interface. Thus, the dendrites may be misaligned or have growth steps on their surfaces so that perfect matching is impossible and dislocations are formed at the interfaces. Dislocations can also form at the interface between crystals of the same orientation but different lattice parameter. The atoms at the interface adjust their positions to give regions of good and bad registry, the latter being *misfit dislocations* (section 9.6). These dislocations are a common feature of solid-state phenomena such as precipitation and epitaxial growth.

The formation of dislocation loops by collapse of vacancy platelets follows directly from previous descriptions of the formation of dislocation loops. (Dislocation loops can also be created by exposing crystals to a flux of high energy atomic particles such as neutrons and ions (see section 3.7), but this process is not considered in this chapter.) In section 1.3 it was shown that, when crystals are rapidly cooled from temperatures close to the melting point, the high-temperature equilibrium concentration of vacancies is retained in a supersaturated state and the vacancies can precipitate to form dislocation loops. A critical supersaturation c_1 is required for the process. This is illustrated by the fact that, in quenched specimens containing a sufficient supersaturation of vacancies to form loops in the centre of grains, no loops are found close to grain boundaries because the degree of super-saturation is reduced by the migration of vacancies to the boundary. The solidification of a melt involves the movement of a liquid–solid interface. There is normally an appreciable temperature gradient in the solid. Immediately below the interface the solid will be very close to the melting point and the equilibrium concentration of vacancies will be high. If this region cools sufficiently rapidly a high density of vacancies will be retained and, providing the supersaturation is greater than c_1, loops will be formed. Speculation about this mechanism revolves around whether or not the supersaturation is sufficient. It is most unlikely to be so when the rate of cooling is slow. However, any loops that are produced in the solid phase below the moving interface will expand by climb due to diffusion of vacancies to the loop. Loops formed at a large angle to the interface may eventually intersect the interface and the two points of emergence formed in this way will act as a site for the propagation of dislocations into the new crystal.

8.3 HOMOGENEOUS NUCLEATION OF DISLOCATIONS

When a dislocation is created in a region of the crystal that is free from any defects the nucleation is referred to as *homogeneous*. This occurs only under extreme conditions because a very large stress is required. A method of estimating the stress has been described by Cottrell (1953). Imagine that in a crystal under an applied resolved shear stress τ, slip is nucleated by the creation on the slip plane of a small dislocation loop of radius r and Burgers vector **b**. The increase in elastic energy of the crystal is given by equation (5.11) and the work done by the applied stress is $\pi r^2 \tau b$. The increase in energy associated with loop creation is therefore

$$E = \frac{1}{2}Gb^2 r \ln\left(\frac{2r}{r_0}\right) - \pi r^2 \tau b \tag{8.1}$$

where ν has been taken as zero for simplicity. The energy increases for small r, reaching a maximum when $dE/dr = 0$ at the critical radius r_c, and then decreases for increasing r. From differentiation of equation (8.1), r_c satisfies the relation

$$r_c = \frac{Gb}{4\pi\tau}\left[\ln\left(\frac{2r_c}{r_0}\right) + 1\right]$$

(8.2)

and the maximum energy is

$$E_c = \frac{1}{4}Gb^2 r_c\left[\ln\left(\frac{2r_c}{r_0}\right) - 1\right]$$

(8.3)

If the loop has at least the critical radius r_c, it will be a stable nucleus and will grow under the applied stress: E_c is the activation energy for the process.

In the absence of thermal fluctuations of energy, nucleation can only occur when $E_c = 0$, i.e. $\ln(2r_c/r_0) = 1$ or $r_c = 1.36r_0$, and from equation (8.2) this requires a stress

$$\tau = Gb/2\pi r_c \simeq G/10$$

(8.4)

in agreement with the theoretical shear stress estimate (equation (1.5)). At the more realistic upper limit for the yield stress of $\tau \simeq G/1000$, equation (8.2) gives $r_c \simeq 500b$ with r_0 taken as $2b$. For a critical nucleus of this size, equation (8.3) gives $E_c \simeq 650Gb^3$, which is about 3 keV for a typical metal. Since the probability of energy E_c being provided by thermal fluctuations is proportional to $\exp(-E_c/kT)$ and $kT = 1/40$ eV at room temperature, it is clear that homogeneous nucleation of dislocations cannot occur at the yield stress. Although stress concentrations in inhomogeneous materials many raise the stress to the level required locally (section 8.4), plastic flow occurs generally by the movement and multiplication of *pre-existing* dislocations.

8.4 NUCLEATION OF DISLOCATIONS AT STRESS CONCENTRATORS

A well-known example of the nucleation of dislocations at a stress concentration is illustrated in Fig. 8.1. Spherical glass inclusions were introduced into a crystal of silver chloride which was subsequently given a treatment to reduce the dislocation density to a low value. The crystal was held at 370°C to homogenize the temperature and remove any internal strains associated with the inclusion, and then cooled to 20°C. The dislocations were revealed by a decoration technique (section 2.3). The photograph shows a row of prismatic dislocation loops, viewed edge on, which have been *punched out* from around the glass inclusion during cooling. The axis of the loops is parallel to a ⟨110⟩ direction, which is the principal slip direction in this structure. The nucleation of the dislocations results from the stress produced

FIGURE 8.1
System of prismatic dislocation loops produced in a recrystallized, dislocation-free crystal of silver chloride to relax the strain field around the small glass sphere caused by differential contraction which occurs during cooling. *(From Mitchell (1958), Growth and Perfection of Crystals, p. 386, Wiley.)*

around the sphere by the differential contraction of the crystal and the glass inclusion during cooling.

Suppose that at 370°C the inclusion has unit radius; it will be resting in a hole in the silver chloride also of unit radius. If the coefficient of expansion of glass and silver chloride are α_1 and α_2 respectively, and $\alpha_1 < \alpha_2$, then on cooling to 20°C the natural radius of the inclusion will be $1 - 350\alpha_1$ and the natural radius of the hole will be $1 - 350\alpha_2$. If the inclusion is unyielding this will result in a spherically symmetrical strain field in the surrounding matrix which can be estimated by analogy with the strain field around a spherical hole with an internal pressure. Thus, consider a particle of radius $r_1(1 + \varepsilon)$ in a hole of natural radius r_1 in an infinite, isotropic elastic medium. The displacement in the matrix is purely radial and decreases with radial distance r as r^{-2}. Since the radial displacement u_r equals εr_1 at $r = r_1$, the displacement at $r (>r_1)$ is

$$u_r = \frac{\varepsilon r_1^3}{r^2} \tag{8.5}$$

The displacement components in rectangular Cartesian coordinates (see section 4.2) are therefore

$$u_x = \varepsilon r_1^3 \frac{x}{r^3} \quad u_y = \varepsilon r_1^3 \frac{y}{r^3} \quad u_z = \varepsilon r_1^3 \frac{z}{r^3} \tag{8.6}$$

where $r^2 = x^2 + y^2 + z^2$. It is easy to show from relations (4.2)–(4.4) that the strain field is one of pure shear, and that the maximum shear strain acting in a radial direction on a cylindrical surface with a radial axis occurs at the interface between the inclusion and matrix and on a cylinder of diameter $\sqrt{2}r_1$. This is illustrated in Fig. 8.2. The magnitude of the corresponding shear stress is

$$\tau_{max} = 3\varepsilon G \tag{8.7}$$

Taking $\alpha_1 = 34 \times 10^{-7} \text{ deg}^{-1}$ for the glass particle and $\alpha_2 = 345 \times 10^{-7} \text{ deg}^{-1}$ for the silver chloride crystal gives $\varepsilon \simeq 0.01$ after the crystal and inclusion have cooled to 20°C and

$$\tau_{max} = \frac{G}{33} \tag{8.8}$$

This is close to the stress required to nucleate dislocations. The dislocations formed in this way at the interface will move away under the influence of the strain field of the inclusion.

FIGURE 8.2

Section of spherical, misfitting particle.

The mechanism for formation of the loops is as follows. Under the action of τ_{max} in the slip direction a small half loop forms on the surface of the glide cylinder indicated in Fig. 8.3. The forward edge component glides away from the interface under the influence of a stress field which decreases in intensity with increasing distance from the interface. The screw components of the loop, which are parallel to the axis of the glide cylinder, experience a tangential force which causes them to glide in opposite directions around the surface of the cylinder. Dislocation movement is restricted to the glide cylinder because any other motion requires climb. When the two ends of the dislocation intersect they annihilate and a dislocation loop will be produced. This is a positive (interstitial) prismatic dislocation loop for $\varepsilon > 0$ and as it

FIGURE 8.3

Mechanism of formation of prismatic loops around an inclusion. (a) Small loop at the surface of inclusion. (b) Loop expands around glide cylinder. (c) Prismatic loop. Burgers vector is parallel to the cylinder axis.

moves away along the slip axis it is effectively moving material away from the inclusion and so relaxing the strain field. The process can be repeated to produce a series of loops. More complex patterns are obtained when glide cylinders of different orientation exist at one particle.

Although this is a somewhat ideal model, dislocation generation at local regions of stress concentration is common. When inclusions and precipitates have a complex shape the associated strain field and resultant dislocation distribution is correspondingly more complex and tangled arrays of dislocations are produced. Dislocation generation also occurs in alloys when precipitates lose coherency (see section 9.7) provided the misfit is sufficiently large ($\varepsilon \simeq 0.01 - 0.05$) for spontaneous nucleation, i.e. zero activation energy. Other stress concentrators such as surface irregularities, cracks, etc., have a similar effect (section 10.10).

8.5 MULTIPLICATION OF DISLOCATIONS BY FRANK–READ SOURCES

To account for the large plastic strain that can be produced in crystals, it is necessary to have *regenerative multiplication* of dislocations. Two mechanisms are important. One is the *Frank–Read type sources* to be described in this section and the other is *multiple cross glide* described in section 8.6.

The first model proposed by Frank and Read resembles the model proposed to account for the role of dislocations in crystal growth. In an irregular array

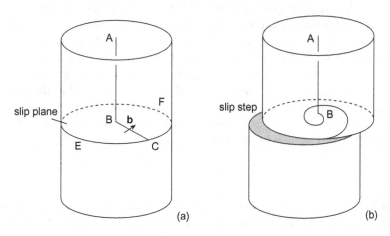

(a)   (b)

FIGURE 8.4
Single ended Frank—Read source. (a) Dislocation lying partly in slip plane *CEF*. (b) Formation of a slip step and spiral dislocation by rotation of *BC* about *B*.

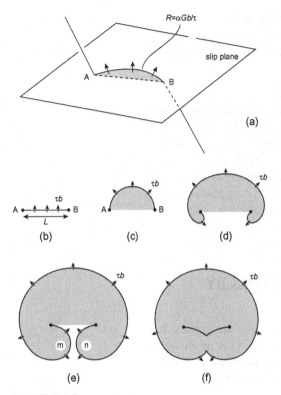

FIGURE 8.5
Diagrammatic
representation of the
dislocation movement in
the Frank–Read source.
Unit slip has occurred in
the shaded area. *(After
Read (1953)*, Dislocations
in Crystals, *McGraw-Hill.)*

some dislocations lie partly in their slip planes and partly in other planes. This is illustrated in Fig. 8.4(a). The length BC of the edge dislocation ABC lies in the slip plane CEF and can move freely in this plane. The length AB is not in the slip plane and is treated as sessile. Thus BC will be anchored at one end and can only move by rotating about B. The dislocation will tend to wind up into a spiral as illustrated in Fig. 8.4(b). Two things are noted about this mechanism:

(a) Each revolution around B produces a displacement b of the crystal above the slip plane relative to that below. The process is *regenerative* since it can repeat itself so that n revolutions will produce a displacement nb. A large *slip step* will be produced at the surface of the crystal.

(b) Spiraling around B results in an increase in the total length of dislocation line. This mechanism is similar to the single ended sources mentioned in section 7.4.

The well-known *Frank–Read source* is an extension of the above mechanism to a dislocation line held at each end as illustrated in Fig. 8.5. The segment AB has the slip plane indicated in Fig. 8.5(a), i.e. its Burgers vector lies in this plane, and is held at both ends by an unspecified barrier, which may be dislocation intersections or nodes, composite jogs, precipitates, etc. An applied resolved shear stress τ exerts a force τb per unit length of line and tends to make the dislocation bow out as described in section 4.4: the radius of curvature R depends on the stress according to equation (4.30). Thus, as τ increases, R decreases and the line bows out until the minimum value of R is reached at the position illustrated in Fig. 8.5(c), where now the slip plane is represented by the plane of the paper. Here, R equals $L/2$, where L is the length of AB, and, with $\alpha = 0.5$ in equation (4.30), the stress is

$$\tau_{max} = Gb/L \qquad (8.9)$$

As the line continues to expand at this stress, R increases and so the dislocation becomes unstable, for equation (4.30) cannot be satisfied unless τ is reduced. The subsequent events are shown in Fig. 8.5(d)–(f). The dislocation forms a large kidney-shaped loop, and the segments m and n annihilate

FIGURE 8.6

Frank–Read source in a silicon crystal. The dislocations have been revealed by the decoration technique described in section 2.3. *(From Dash (1957)*, Dislocation and Mechanical Properties of Crystals, *Wiley.)*

on meeting. This occurs because m and n, which move in opposite directions under the same stress, have the same Burgers vector but opposite line sense. The result is a large outer loop, which continues to expand, and a regenerated dislocation AB, which repeats the process.

Figure 8.5 and equation (4.30) neglect the orientation of the Burgers vector in the slip plane: as explained in section 4.5, the bowing line usually adopts in practice a shape elongated in the Burgers vector direction. An excellent example of a Frank–Read source is illustrated in Fig. 8.6. The dislocation is held at each end by other parts of the dislocation network. These are not in the {111} plane of the loops and are consequently out of focus. The dislocation lines tend to lie along the ⟨110⟩ directions in which they have a minimum energy, indicating the strong anisotropy of the dislocation core energy of this material.

The expression for $\tau_{\max}$ shows that dislocation segments of length $L \simeq 10^4 b$ can act as Frank–Read sources at applied stresses close to the yield stress. Considerable multiplication probably occurs by the Frank–Read mechanism, but additional processes must occur to account for the experimental observations. For example, the Frank–Read source does not explain the broadening of slip bands which is a common feature of the early stages of deformation of some crystals, e.g. iron and lithium fluoride, and is described in the following.

FIGURE 8.7

Lateral growth of glide bands in lithium fluoride. (a) Glide band formed at a single loop. Large pits show the position of the loop. Small pits show that new dislocations lie on both sides of the glide plane of the original loop. (b) Widening of a glide band. *(From Gilman and Johnston,* Solid State Physics *13, 147, 1962.)*

8.6 MULTIPLICATION BY MULTIPLE CROSS GLIDE

An example of the multiplication of dislocations associated with the initiation and broadening of a slip band is shown in Fig. 8.7, from work by Gilman and Johnston on lithium fluoride. The dislocations are revealed by the etch pit technique (section 2.3); a succession of etching and deformation treatments revealed the growth of the slip band. Initially (Fig. 8.7(a)), the slip band started as a single loop of dislocation and the two points of emergence are revealed by the two large pits. Deformation resulted in the formation of dislocations on each side of the original loop. Since the band of dislocation etch pits has a finite width, it follows that the dislocations do not lie on the same glide plane, but on a set of parallel glide planes. Widening of a glide band is shown in Fig. 8.7(b); after the first deformation the band was W_1 wide and after the second deformation W_2 wide. The dislocation density is approximately uniform throughout the band.

The widening can be accounted for by a process called *multiple cross glide*. In principle, it is the same as the process illustrated in Fig. 3.9. Thus, a screw dislocation lying along *AB* can cross glide onto position *CD* on a parallel glide plane. If the stress is greater on the primary plane the long jogs *AC* and *BD* are relatively immobile. However, the segments lying in the primary slip planes will be free to expand and can each operate as a Frank–Read source. A similar mechanism can result from cross-slipped segments in particle strengthened alloys (Fig. 7.12). When cross glide can occur readily, the Frank–Read sources may never complete a cycle and there will be one continuous dislocation line lying on each of many parallel glide planes, connected by jogs. Thus, it is possible for a single dislocation loop to expand and multiply in such a way that the slip spreads from plane to plane, so producing a wide slip band. Multiple cross glide is a more effective mechanism than the simple Frank–Read source since it results in a more rapid multiplication.

8.7 MULTIPLICATION BY CLIMB

Two mechanisms, involving climb, which increase the total dislocation length, have been described already; (a) the expansion of a prismatic loop and (b) the spiraling of a dislocation with a predominantly screw character (section 3.7). A *regenerative multiplication* known as the *Bardeen–Herring source* can occur by climb in a similar way to the Frank–Read mechanism illustrated in Fig. 8.5. Suppose that the dislocation line *AB* in Fig. 8.5 is an edge dislocation with Burgers vector perpendicular to the plane of the paper in Fig. 8.5(b)–(f), i.e. the plane shown in Fig. 8.5(a) is not the slip plane of *AB* but contains its extra half-plane. If the dislocation line is held at *A* and *B*, the presence of an excess concentration of vacancies will cause the dislocation to climb, the force causing climb being given by equation (4.40). An additional condition to the normal Frank–Read source must be satisfied if this process is to be regenerative; the anchor points *A* and *B* must end on dislocations with screw character. If this were not so, one cycle of the source, Fig. 8.5(e), would remove the extra half-plane without creating a new dislocation along *AB*. Bardeen–Herring sources have been observed experimentally, as illustrated in Fig. 8.8 for an aluminum–3.5 per cent magnesium alloy quenched from 550°C into silicone oil. The large concentration of excess vacancies has resulted in the formation of four concentric loops. Each loop represents the removal of a plane of atoms. More complicated source arrangements have also been observed in this and other alloys (see 'Further Reading').

0.5 μm

FIGURE 8.8

Transmission electron micrograph of concentric loops formed at a climb source in aluminum—3.5 per cent magnesium alloy quenched from 550°C. *(From Westmacott, Barnes and Smallman,* Phil. Mag. **7**, *1585, 1962.)*

For a dislocation line bowing by slip, the work done by the applied shear stress balances in equilibrium the increase in line energy (section 4.5). Similarly for bowing by climb, the increase in line energy is balanced by the gain in energy due to the loss or creation of point defects (section 4.7). Hence, from equations (4.30) and (4.40), the chemical force per unit length which produces a radius of curvature R by climb is

$$f = \frac{bkT}{\Omega} \ln\left(\frac{c}{c_0}\right) = \frac{\alpha Gb^2}{R} \tag{8.10}$$

By analogy with the analysis for the Frank–Read source, the critical vacancy supersaturation required to operate a Bardeen–Herring source of length L is therefore

$$\ln\left(\frac{c}{c_0}\right) = \frac{2\alpha Gb\Omega}{LkT} \tag{8.11}$$

where $\alpha \simeq 0.5$. Consider a typical metal with $\Omega = b^3$ and $Gb^3 = 5$ eV. At $T = 600$ K, $kT \simeq 0.05$ eV and equation (8.11) gives $\ln(c/c_0) \simeq 100b/L$. Thus, for $L = 10^3 b$, $c/c_0 = 1.11$, and for $L = 10^4 b$, $c/c_0 = 1.01$. These are small supersaturations in comparison with those met in rapid quenches, for which c/c_0 can be as large as $\sim 10^4$, and Bardeen–Herring sources can probably operate throughout the period of fast cooling.

8.8 GRAIN BOUNDARY SOURCES

An important source of dislocations during plastic flow in poly-crystalline materials is the boundary region between the grains. Many investigations have shown that dislocations can be emitted from grain boundaries, and it is probable that they subsequently multiply within grains by the multiple cross-slip process (section 8.6). Several mechanisms may be involved in the emission. In low-angle boundaries (Chapter 9), the segments of dislocation networks forming the boundary can act as Frank–Read (or Bardeen–Herring) sources. Another possible source is an adsorbed lattice dislocation which is stabilized at the boundary in the stress-free state by the boundary structure. The resulting boundary ledge can provide sites for dislocation nucleation under stress, for the reduction in boundary area at nucleation reduces the activation energy. The dislocations generated by sources within a grain produce large stress concentrations when piled up at the grain boundary (section 10.9): these can trigger boundary sources at relatively low applied stress. It has also been established that migrating grain boundaries generate dislocations in the lattice they pass through. Gleiter *et al.* (1980) have proposed that these dislocations are nucleated by accidental mispacking of atoms at the boundary when one grain grows at the expense of another. This process provides a source for dislocations in crystals grown in the solid state, such as those formed by recrystallization and phase transformations.

FURTHER READING

Ashby MF, Johnson L: On the generation of dislocations at misfitting particles in a ductile matrix, *Phil Mag* 20:1009, 1969.

Boyd JD, Edington JW: Dislocation climb sources and vacancy loops in quenched Al-2.5%Cu, *Phil Mag* 23:633, 1971.

Brown LM, Woolhouse GR: The loss of coherency of precipitates and the generation of dislocations, *Phil Mag* 21:329, 1970.

Cottrell AH: *Dislocations and plastic flow in crystals*, 1953, Oxford University Press.

Elbaum C: Substructure in crystals grown from the melt, *Progress in metal physics*, vol 8, 1959, Pergamon, p. 203.

Frank FC: Crystal growth and dislocations, *Adv Physics* 1:91, 1952.

Gilman JJ, Johnston WG: Origin and growth of glide bands in LiF crystals, in Fisher JC et al., editors: *Dislocations and mechanical properties of crystals*, 1957, Wiley, p. 116.

Gleiter H, Mahajan S, Bachman KJ: The generation of lattice dislocations by migrating boundaries, *Acta Met* 28:1603, 1980.

Hirth JP, Lothe J: *Theory of dislocations*, 1982, Wiley.

Li JCM: Dislocation sources, in Ashby MF, editor: *Dislocation modelling of physical systems*, 1981, Pergamon, p. 498.

Mutaftschiev B: Crystal growth and dislocations, in Nabarco FRN, editor: *Dislocations in solids*, vol 5, 1980, North-Holland, p. 57.

Pamplin BR, editor: *Crystal growth*, 1975, Pergamon.

CHAPTER 9

Dislocation Arrays and Crystal Boundaries

9.1 PLASTIC DEFORMATION, RECOVERY AND RECRYSTALLIZATION

Plastic deformation of crystalline materials leads to the formation of three-dimensional arrays or distributions of dislocations which are characteristic of (1) crystal structure of the material being deformed, (2) temperature of deformation, (3) strain and (4) strain rate. Additionally, such features as grain boundaries, precipitates, and stacking fault energy affect the distribution of the dislocations. Two distributions are illustrated in Fig. 9.1, which shows the effect of deforming pure iron specimens at 20°C and −135°C, respectively. Deformation at 20°C has resulted in the formation of dense *tangles of dislocations* arranged in *walls* surrounding regions or *cells* almost free from dislocations. The *cell size* reaches a limit in the early stages of deformation and changes only slightly thereafter. The cell walls tend to orient themselves in certain crystallographic directions. Deformation at −135°C produces a much more homogeneous distribution of dislocations.

The hardening of crystals during plastic deformation is due to the increase in dislocation density and the mutual interaction between dislocations (section 10.8). Most of the work done by the external load during plastic deformation is dissipated as heat, but a small proportion is retained in the material as *stored energy*. This is accomplished by an increase in elastic strain energy resulting from the increase in dislocation density. An additional, small amount of energy is stored when point defects are produced during plastic deformation. The energy only remains stored within a crystal if the temperature is sufficiently low for the atoms to be effectively immobile, i.e. $T \lesssim 0.3\ T_m$: the plastic deformation that meets this requirement is known as *cold-work*. The stored energy can be released if the dislocations rearrange themselves into configurations of lower energy. These are called low-angle boundaries and can be represented by a uniform array of one, two, three or more sets of dislocations (sections 9.2 and 9.3). They separate regions of the

171

Introduction to Dislocations.

FIGURE 9.1
Dislocation arrangements produced by plastic deformation of iron. (a) Dense tangles and dislocation free cells formed after 9 per cent strain at 20°C. (b) Uniform array of straight dislocations formed after 7 per cent strain at −135°C. *(From Keh and Weissmann (1963)*, Electron Microscopy and Strength of Crystals, *p. 231, Interscience.)*

crystal which differ in orientation by ≲5°. A considerable amount of energy is released by the local rearrangement of the dislocations in the *tangles* and further release of energy occurs when low-angle boundaries are formed. Both these processes involve climb of the dislocations and will occur, therefore, only when there is sufficient thermal activation to allow local and long-range diffusion of point defects, i.e. $T \gtrsim 0.3\ T_m$. These changes are accompanied by a pronounced softening of the dislocation-hardened crystal. The process is called *recovery* and it occurs when a plastically deformed crystal is heated to moderate temperatures. The later stages of the recovery process in which low-angle boundaries are formed is called *polygonization*.

When a heavily cold-worked metal is heated above a critical temperature new grains with relatively low dislocation density are produced in the 'recovered' structure, resulting in a process called *recrystallization. Large-angle grain boundaries* with a misorientation ≳10° are produced. The sequence of photographs in Fig. 9.2 illustrates the change from a heavily deformed structure with a uniform distribution of tangled dislocations to a recrystallized structure containing large and small angle grain boundaries. The grain structure is small immediately after recrystallization but grows progressively with longer annealing times and higher temperatures. This is called *grain growth* and results in a small reduction in energy because the total area of grain boundary is reduced. Some recent advances in our knowledge of the dislocation content of high-angle grain boundaries are described in section 9.6.

The basic process in the formation of low-angle boundaries is illustrated in its most simple form in Fig. 9.3. Consider a crystal which is bent about its

FIGURE 9.2

Transmission electron micrographs illustrating the structure of deformed and annealed 3.25 per cent silicon iron. (a) Approximately uniform distribution of dislocations in a crystal rolled 20 per cent. (b) Formation of small sub-grains in rolled material annealed 15 min at 500°C. (c) As for (b) annealed 15 min at 600°C. (d) As for (b) annealed 30 min at 600°C. *(From Hu, Trans. Met. Soc. AIME 230, 572, 1964.)*

z-axis: the resulting dislocations will be distributed randomly on the glide planes and there will be an excess of edge dislocations of one sign, as illustrated in Fig. 9.3(a). Excess dislocations of one sign, formed in this case as a result of the strain gradient due to bending, are known as *geometrically necessary dislocations*. The energy of the crystal can be reduced by rearranging the dislocations into a vertical wall to form a symmetrical tilt boundary, Fig. 9.3(b) (section 9.2). Alternatively, imagine that a thin wedge-shaped section, symmetrical about the plane $x = 0$, is cut from the perfect crystal *ABCD*, Fig. 9.4 (a), and that the two cut faces are then placed together as in Fig. 9.4(b). This is geometrically the same as the tilt boundary in Fig. 9.3(b) and illustrates an important feature of small angle boundaries, namely that they have

FIGURE 9.3

Formation of a low-angle boundary. The orientation of the slip planes is denoted by a cross. (a) Bent crystal with random dislocations. (b) Rearrangement of dislocations to form a symmetrical tilt boundary. Both climb and glide are required to produce this boundary.

FIGURE 9.4

Geometry of the formation of a symmetrical tilt boundary. The relative rotation of the two crystals is produced by cutting a wedge from the perfect crystal.

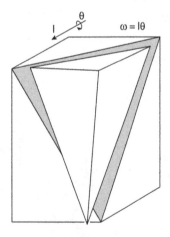

FIGURE 9.5

Formation of a general grain boundary with five degrees of freedom. Three degrees arise from the rotation about **l** of one grain with respect to the other: vector **l** has three components.

no long range stress field. The strain is localized in the region around the dislocations.

The example illustrated in Fig. 9.4 is a boundary with *one degree of freedom* since the axis of relative rotation is fixed in the same crystallographic direction in both crystals and the boundary is a mirror plane of the bicrystal. In the general case illustrated in Fig. 9.5, the boundary is formed by cutting a single crystal on two arbitrary planes, removing the wedge of material between the cuts, and placing the grains together with a twist. The boundary has *five* degrees of freedom, three corresponding to the rotation of one grain with respect to the other about three perpendicular axes, and two due to the rotation of the boundary with respect to the two grains individually (see section 9.3).

The requirement that there are no long-range stress fields places severe conditions on the dislocation geometry of low-angle boundaries and this is discussed in sections 9.3 and 9.4. The regular dislocation arrays observed after recovery are true low-angle boundaries. They can adopt the form typified in Fig. 9.3 because there are no external constraints on the two grains. These boundaries are unlikely to form during plastic deformation, however, because the material in regions away from the boundary will be unable to adopt, in general, the orientation required: long-range strain fields then result. The tangled arrangements in Fig. 9.1 may well be of this form.

9.2 SIMPLE DISLOCATION BOUNDARIES

The simplest boundary is the *symmetrical tilt boundary* (Figs 2.12(b), 9.3 and 9.4). The atomic mismatch at the boundary is accommodated by regions of good and bad fit: the latter are dislocations. In a simple cubic lattice with edge dislocations $\mathbf{b} = [100]$, the boundary will consist of a sheet of equally spaced dislocations lying parallel to the z-axis; the plane of the sheet will be the symmetry plane $x = 0$, i.e. (100). The extra half-planes of the dislocations terminate at the boundary from the left- and right-hand sides alternately. The crystals on each side of the boundary are rotated by equal and opposite amounts about the z-axis and differ in orientation by the angle θ (Fig. 2.12(b)). If the spacing of the dislocations is D, then:

$$\frac{b}{2D} = \sin\frac{\theta}{2} \qquad (9.1)$$

and for small values of θ (in radians)

$$\frac{b}{D} \sim \theta \qquad (9.2)$$

If $\theta = 1°$ and $b = 0.25$ nm the spacing between dislocations will be 14 nm. When the spacing is less than a few lattice spacings, say five, $\theta \sim 10°$ and the individual identity of the dislocations is in doubt, the boundary is called a *large-angle boundary*. Experimental evidence for tilt boundaries has been presented in Fig. 2.12. Many such observations of boundaries have been reported using decoration and thin film transmission electron microscope techniques.

A *more general tilt boundary* is illustrated in Fig. 9.6. By introducing a second degree of freedom the boundary plane can be rotated about the z-axis. The boundary plane is no longer a symmetry plane for the two crystals but forms an angle ϕ with the mean [100] direction of the two grains. Two sets of atom planes end at the boundary and the geometrical conditions require the boundary to consist of two sets of uniformly spaced edge dislocations with mutually perpendicular Burgers vectors, $\mathbf{b}_1$ and $\mathbf{b}_2$. Since $EC = AC \cos(\phi - \theta/2)$ and $AB = AC$

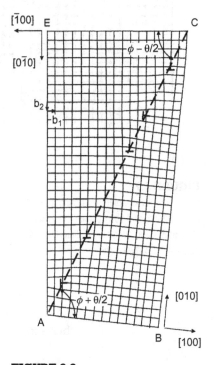

FIGURE 9.6

Tilt boundary with two degrees of freedom. This is the same boundary as in Fig. 9.3 except that the plane of the boundary makes an arbitrary angle ϕ with the mean of the (010) planes in the two grains. The atom planes end on the boundary in two sets of evenly spaced edge dislocations. *(After Read (1953), Dislocations in Crystals, McGraw-Hill.)*

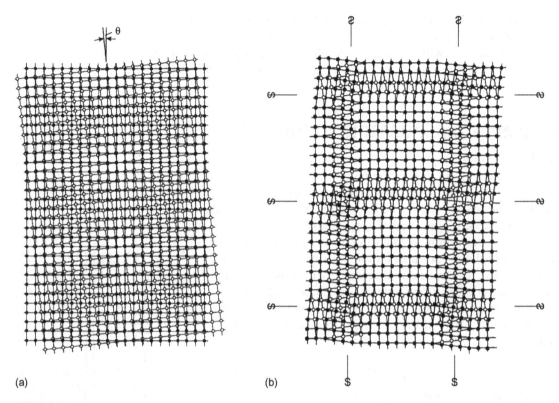

(a) (b)

FIGURE 9.7
Formation of a simple twist boundary by rotation about an axis perpendicular to the figure: open circles represent atoms just above the boundary and solid circles those just below. (a) Atom positions resulting from rigid rotation through angle θ. (b) Accommodation of mismatch in (a) by localized distortions, seen to be two sets of parallel screw dislocations labeled S-S. *(After Read (1953) Dislocations in Crystals, McGraw-Hill.)*

$\cos(\phi + \theta/2)$, the number of (100) planes intersecting EC is $(AC/b_1)\cos(\phi - \theta/2)$ and the number intersecting AB is $(AC/b_1)\cos(\phi + \theta/2)$. The number terminating at the boundary is therefore

$$(AC/b_1)[\cos(\phi - \theta/2) - \cos(\phi + \theta/2)] \simeq (AC/b_1)\theta \sin \phi$$

if $\theta \ll 1$. Similarly, the number of terminating (010) planes is $(AC/b_2)\theta \cos \phi$. The spacings along the boundary of the two sets of dislocations are therefore

$$D_1 = \frac{b_1}{\theta \sin \phi} \quad D_2 = \frac{b_2}{\theta \cos \phi} \tag{9.3}$$

A simple boundary formed from a *cross grid of pure screw dislocations* is illustrated in Fig. 9.7. A single set of screw dislocations has a long-range stress field and is therefore unstable but the stress field is cancelled by the second set of screw dislocations. The two sets of equally spaced parallel dislocations lie in the boundary which also lies in the plane of the diagram. They produce a rotation about an axis normal to the boundary of one-half of the crystal with respect to the other. Such a boundary is called a *twist boundary*. The spacing between dislocations in each set is:

$$D = \frac{b}{\theta}$$

9.3 GENERAL LOW-ANGLE BOUNDARIES

It was explained in section 9.1 that the general planar boundary has five degrees of freedom. Three correspond to the difference in orientation of the grains, for if $\mathbf{l}$ is a unit vector parallel to the axis of relative rotation of the grains (Fig. 9.5), the rotation may be represented by the vector $\omega = \mathbf{l}\theta$, which has three independent components. The orientation of the boundary may be specified by a unit vector $\mathbf{n}$ normal to the boundary plane. This introduces only two degrees of freedom, one for each crystal, for rotation of the boundary about $\mathbf{n}$ is not significant. The tilt and twist boundaries can now be defined as follows: in a *pure tilt* boundary, $\mathbf{l}$ and $\mathbf{n}$ are at right-angles and in a *pure twist* boundary $\mathbf{l} = \mathbf{n}$. In the examples in section 9.2 the tilt boundaries are described by arrays of edge dislocations and the twist boundaries by screw dislocations, but these are special cases. For example, if the dislocations in the crystals of Fig. 2.12(b) have Burgers vectors $\frac{1}{2}\langle 111 \rangle$ rather than $\langle 100 \rangle$, the pure tilt boundary on (100) would consist of a wall of dislocations of mixed character, the edge components all having the same sign and the screw components alternating in sign; e.g. the Burgers vectors may alternate between $\frac{1}{2}[111]$ and $\frac{1}{2}[1\overline{1}\overline{1}]$. The screw components contribute to the energy of the tilt boundary but not to its misfit. Similarly, the pure twist boundary will, in general, contain dislocations with (alternating) edge character. The general boundary with $\mathbf{l}$ neither perpendicular nor parallel to $\mathbf{n}$ has *mixed* tilt and twist character.

The orientation of the crystals and the boundary formed in the way illustrated in Figs 9.4 and 9.5 can be defined by the parameters $\mathbf{l}$, $\mathbf{n}$ and θ. Frank (1950) has derived a relation which may be used to determine the arrangements of dislocations which will produce such a boundary, or conversely the orientation of a boundary produced by a given set of dislocations or *dislocation network*. For a general boundary, Frank's relation is

$$\mathbf{d} = (\mathbf{r} \times \mathbf{l})2\sin\frac{\theta}{2}$$

or   (9.4)

$$\mathbf{d} = (\mathbf{r} \times \mathbf{l})\theta$$

for small values of θ. The vector $\mathbf{r}$ represents an arbitrary vector lying in the plane of the boundary which contains the dislocation network, and $\mathbf{d}$ is the sum of the Burgers vectors of all the dislocations intersected by $\mathbf{r}$, i.e.

$$\mathbf{d} = \sum_i N_i \mathbf{b}_i \qquad (9.5)$$

where N_i is the number of dislocations of Burgers vectors $\mathbf{b}_i$ cut by $\mathbf{r}$. Relation (9.4) is derived by application of the Burgers circuit construction (section 1.4). Consider a closed circuit starting at the end point of $\mathbf{r}$, passing through one grain to the start point of $\mathbf{r}$ and returning back through the second grain to the start point of $\mathbf{r}$ and returning back through the second grain. In a perfect reference crystal, the circuit starts at the end point of $\mathbf{r}$ and finishes at the end point of the vector $\mathbf{r}'$ obtained from $\mathbf{r}$ by the rotation $\omega = \mathbf{l}\theta$. Vector $\mathbf{d}$ equals the closure failure $\mathbf{r} - \mathbf{r}'$ of the circuit which, for small θ, equals $\mathbf{r} \times \omega = (\mathbf{r} \times \mathbf{l})\theta$.

A number of points should be noted regarding this relation. (a) It applies only to boundaries which are essentially flat and have no long-range stress field, i.e. the elastic distortion is restricted to the region close to the dislocations. (b) The formula does not uniquely determine the dislocations present, or their pattern, for a given orientation of crystal and boundary. Thus, a variety of possibilities may arise and the most probable will be the one of lowest energy. (c) The density of a given set of dislocations in a boundary is directly proportional to θ (for small θ). (d) Each set of dislocation lines will be straight, equally spaced and parallel even for a boundary containing several sets of dislocations with different Burgers vectors.

A general boundary requires three sets of dislocations with three non-coplanar Burgers vectors, so that the boundaries formed from one or two sets are restricted. Frank's formula can be applied to analyse all cases, either to determine the possible dislocation arrangements if $\mathbf{n}$, $\mathbf{l}$ and θ are known or to find the orientation, etc. if the dislocation content is specified. This is illustrated here by simple examples. (Examples of more detailed cases may be found in Chap. 19 of Hirth and Lothe, 1982.)

If a boundary contains only *one* set of dislocations, each having Burgers vector $\mathbf{b}$, equations (9.4) and (9.5) become

$$N\mathbf{b} = (\mathbf{r} \times \mathbf{l})\theta \qquad (9.6)$$

Clearly $\mathbf{b}$ is perpendicular to $\mathbf{r}$ and $\mathbf{l}$, and since the direction of $\mathbf{r}$ in the boundary is arbitrary, $\mathbf{b}$ is parallel to $\mathbf{n}$ and $\mathbf{l}$ is perpendicular to $\mathbf{n}$. When $\mathbf{r}$ is chosen parallel to $\mathbf{l}$, $\mathbf{r} \times \mathbf{l} = 0$ and no dislocations are intersected by $\mathbf{r}$, showing that the dislocations are parallel to the rotation axis $\mathbf{l}$. From these observations, the boundary is a pure tilt boundary consisting of pure edge dislocations. If **r** is taken perpendicular to **l** and (obviously) **n**, then **r** = (**l** × **n**)r, where $r = |$**r**$|$ is the length of **r**. Equation (9.6) then becomes

$$N\mathbf{b} = [(\mathbf{l} \times \mathbf{n}) \times \mathbf{l}]r\theta$$

i.e.

$$N\mathbf{b} = \mathbf{n}r\theta$$

Since **b** = b**n**, the spacing of the dislocations is

$$D = \frac{r}{N} = \frac{b}{\theta}$$

in agreement with equation (9.2). The simple tilt boundary of Figs 9.4 and 2.12(b) with **l** = [001] and **n** = [100] is of this type, for when **r** lies in the [001] direction, **r** × **l** is zero, demonstrating that the dislocations are all parallel to [001]. When **r** is in the [010] direction **r** × **l** equals r[100], showing that if the dislocations have Burgers vectors **b** = [100], their spacing is b/θ.

Consider next a boundary containing *two* sets of dislocations with Burgers vectors **b**$_1$ and **b**$_2$, which are not parallel. Frank's formula is

$$N_1\mathbf{b}_1 + N_2\mathbf{b}_2 = (\mathbf{r} \times \mathbf{l})\theta \tag{9.7}$$

Taking the scalar product of both sides with **b**$_1$ × **b**$_2$ gives

$$(\mathbf{r} \times \mathbf{l}) \cdot (\mathbf{b}_1 \times \mathbf{b}_2) = 0$$

i.e.

$$\mathbf{r}\cdot[\mathbf{l} \times (\mathbf{b}_1 \times \mathbf{b}_2)] = 0 \tag{9.8}$$

Since the direction of **r** in the boundary is arbitrary, this condition is satisfied whenever [**l** × (**b**$_1$ × **b**$_2$)] is parallel to **n**; i.e. (**b**$_1$ × **b**$_2$) and **l** lie in the boundary plane. This defines a pure tilt boundary. With **r** chosen parallel to **l**, the right-hand side of equation (9.7) is zero and so both sets of dislocations are parallel to **l**. Their density is found by setting **r** perpendicular to **l**, i.e. **r** = (**l** × **n**)r, and equation (9.7) becomes

$$N_1\mathbf{b}_1 + N_2\mathbf{b}_2 = \mathbf{n}r\theta \tag{9.9}$$

The boundary of Fig. 9.6 has this form. From the figure, **b**$_1$ = [100], **b**$_2$ = [0$\bar{1}$0] and **n** = [sin ϕ, −cos ϕ, 0]. Substitution into equation (9.9) gives $D_1 = b/\theta$ sin ϕ and $D_2 = b/\theta$ cos ϕ, where b is the lattice parameter, in agreement with the result (9.3).

Condition (9.8) is also satisfied when **l** is parallel to **b**$_1$ × **b**$_2$, i.e. **l** = (**b**$_1$ × **b**$_2$)/|**b**$_1$ × **b**$_2$|. Equation (9.7) then becomes

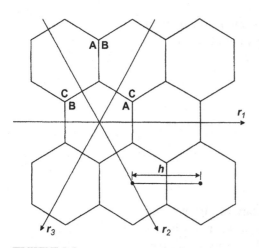

FIGURE 9.8

Hexagonal network of perfect dislocations ($\mathbf{b} = \frac{1}{2}\langle 110\rangle$) on a {111} plane in a face-centered cubic lattice.

$$N_1\mathbf{b}_1 + N_2\mathbf{b}_2 = [\mathbf{b}_1(\mathbf{r} \cdot \mathbf{b}_2) - \mathbf{b}_2(\mathbf{r} \cdot \mathbf{b}_1)]\theta/|\mathbf{b}_1 \times \mathbf{b}_2|$$
(9.10)

It may be shown that the two sets of dislocations are not parallel. For the pure twist boundary, $\mathbf{n}$ equals $\mathbf{l}$ and the spacings simplify to $\mathbf{b}_1/\theta$ and $\mathbf{b}_2/\theta$ respectively. In the special case when $\mathbf{b}_1$ is perpendicular to $\mathbf{b}_2$, the line directions are $\mathbf{b}_1$ and $\mathbf{b}_2$, respectively, so that the dislocations are pure screw. This corresponds to the boundary of Fig. 9.7.

Finally, consider the following boundary in a face-centered cubic lattice. The dislocation network is an example of that defined by equation (9.10), except that at the point where a dislocation of one set intersects one from the other, they react to form a third. This results in an hexagonal net of lower line length (and energy), as illustrated in Fig. 9.8. The Burgers vectors are described using the Thompson notation (section 5.4). They lie in the plane of the net, e.g. $\mathbf{AB} = \frac{1}{2}[\bar{1}10]$, $\mathbf{BC} = \frac{1}{2}[10\bar{1}]$, $\mathbf{CA} = \frac{1}{2}[0\bar{1}1]$ in plane (111), and one results from the addition of the other two; e.g. $\mathbf{AB} + \mathbf{BC}$ equals $\mathbf{AC}$. It is important that the sequence of lettering around any node is consistent. The Burgers vector of each dislocation is represented by two letters written on either side of it. The lettering should be such that any track around the node makes a continuous path, for example, AB, BC, CA, or AB, CA, BC, but not BC, CA, BA, or AB, BC, AC. Given that h is the distance between the centers of the mesh, the orientation of the plane of the net and the individual dislocations in the net can now be determined. If vector $\mathbf{r}$ is placed horizontally on the net in Fig. 9.8, i.e. $\mathbf{r}_1$, it crosses r_1/h dislocations of Burgers vector $\mathbf{AB}$. Frank's formula becomes

$$\mathbf{AB}r_1/h = (\mathbf{r}_1 \times \mathbf{l})\theta$$

Similarly for vectors $\mathbf{r}_2$ and $\mathbf{r}_3$ shown in the figure:

$$\mathbf{BC}r_2/h = (\mathbf{r}_2 \times \mathbf{l})\theta$$
$$\mathbf{CA}r_3/h = (\mathbf{r}_3 \times \mathbf{l})\theta$$

It follows that $\mathbf{l}$ is perpendicular to the net, which therefore forms a pure twist boundary. Furthermore, since $\mathbf{AB}$ is perpendicular to $\mathbf{r}_1$ and $\mathbf{l}$, the dislocation segments perpendicular to $\mathbf{r}_1$ are pure screw in character. So, too, are those perpendicular to $\mathbf{r}_2$ and $\mathbf{r}_3$. All the dislocations are pure screws, and θ, the angle of rotation of the boundary, is b/h for small θ, where $b = |\mathbf{AB}|$. Such a net of regular hexagons cannot lie in any other plane without producing long-range stress fields in the crystal. Any deviation from the (111) boundary plane must be accompanied by a modification of the regular hexagon structure if long-range stresses are to be avoided. The dislocations can dissociate to form extended nodes (section 7.8) whilst still satisfying Frank's formula. Similar stable networks can also arise in other structures. The arrangement of dislocations in a typical boundary in a body-centered cubic crystal is shown in Fig. 9.9.

Although low-angle boundaries are formed primarily under conditions in which dislocations can climb freely, it is possible to produce them by slip, but the geometrical conditions are very restrictive. The orientation and character of the dislocations in the boundary must have common slip Burgers vectors and lie on slip planes, and also satisfy the Frank equation.

9.4 STRESS FIELD OF DISLOCATION ARRAYS

Dislocation nets may have long- or short-range stress fields. The distribution of stress is sensitive to the arrangement, orientation and Burgers vectors of the dislocations. Boundaries satisfying Frank's condition (equation (9.4)) do not produce long-range fields. A few examples of the most elementary boundaries are sufficient to illustrate these points.

The components of the stress field of single edge and screw dislocations have been given in equations (4.16) and (4.13), respectively. The total stress field of an array is obtained by a summation of the components of the stress field of the individual dislocations sited in the array. Thus, consider a wall of edge dislocations of Burgers vector **b** making up a symmetrical tilt boundary lying in the plane $x = 0$ with the dislocations parallel to the z-axis. With one dislocation lying along the z-axis, and N others above and N below, the stress field of the boundary is given by

FIGURE 9.9

Transmission electron micrograph of extensive dislocation networks in body-centered cubic iron. Each network consists of three sets of dislocations, Burgers vectors $\frac{1}{2}[111]$, $\frac{1}{2}[1\overline{1}\overline{1}]$ and [100]. The plane of the networks is almost parallel to the plane of the foil. (Courtesy Dadian and Talbot-Besnard.)

$$\sigma_{xx} = \frac{-Gb}{2\pi(1-\nu)} \sum_{n=-N}^{N} \frac{y_n(3x^2 + y_n^2)}{(x^2 + y_n^2)^2} \qquad (9.11)$$

$$\sigma_{yy} = \frac{Gb}{2\pi(1-\nu)} \sum_{n=-N}^{N} \frac{y_n(x^2 - y_n^2)}{(x^2 + y_n^2)^2} \qquad (9.12)$$

$$\sigma_{xy} = \frac{Gb}{2\pi(1-\nu)} \sum_{n=-N}^{N} \frac{x(x^2 - y_n^2)}{(x^2 + y_n^2)^2} \qquad (9.13)$$

where $y_n = y - nD$ and D is the dislocation spacing. The term with $n = 0$ is the contribution of the dislocation at $y = 0$ (see equation (4.16)). The effect

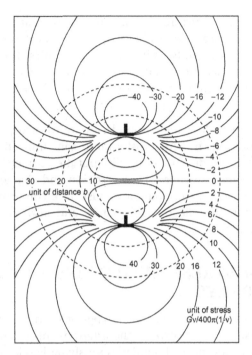

FIGURE 9.10
Constant value contours of the stress field σ_{zz} of a single edge
dislocation. Dashed circles show distance from the origins. *(From
Li (1963), Electron Microscopy and Strength of Crystals, p. 713.
Interscience.)*

FIGURE 9.11
The stress field σ_{zz} of a vertical wall of two edge dislocations
spaced $28b$ apart. *(From Li (1963), Electron Microscopy and
Strength of Crystals, p. 173. Interscience.)*

of the summation is for the individual contributions to tend to cancel within
a distance $\sim ND$ of the origin. For example, it is readily shown from the
above equations that the stresses, σ_{xx}, σ_{yy} and $\sigma_{zz} = -\nu(\sigma_{xx} + \sigma_{yy})$ at $y = D/2$
due to dislocation $n = 0$ are cancelled by those due to dislocation $n = 1$. This
mutual cancellation occurs over a large part of the region $0 < y < D$, as demonstrated explicitly for σ_{zz} by comparing Figs 9.10 and 9.11. Addition of dislocations $n = -1$, $n = 2$, and then $n = -2$, $n = 3$, etc., extends the region of
cancellation. Exact summation of equations (9.11)–(9.13) is possible for
infinite arrays, i.e. $N \to \infty$. The solution for σ_{xy} is shown in Fig. 9.12. For,
$|x| \gtrsim D/2\pi$, σ_{xy} is approximated by

$$\sigma_{xy} \simeq \frac{2\pi Gbx}{(1-\nu)D^2} \cos\left(\frac{2\pi y}{D}\right) \exp\left(\frac{-2\pi|x|}{D}\right) \tag{9.14}$$

so that away from the boundary, σ_{xy} decreases exponentially with x. Near the
wall ($|x| \lesssim D/2\pi$), the stress is dominated by the nearest one to three

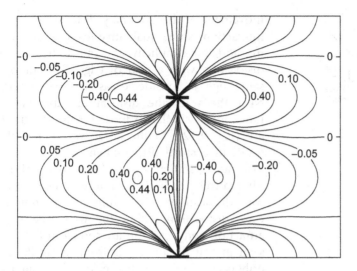

FIGURE 9.12
The shear stress field σ_{xy} of an infinite array of edge dislocations. Unit of stress $Gb/2(1-\nu)D$. (From Li, Acta Metall. 8, 296, 1960.)

dislocations (Fig. 9.12). Because the stress fields are localized at the boundary the strain is small and therefore the boundary represents a stable configuration with respect to slip. However, the dislocations in the boundary can climb, resulting in an increased separation of the dislocations and a reduction in θ. Although the infinite wall of edge dislocations has no long-range stress field, a finite wall (N not infinite) has, as implied by Fig. 9.11.

If the dislocations in the vertical wall were to be moved by slip so that the boundary made an angle ϕ with the original low-energy position, it can be shown that the wall will have a long-range stress field and a higher energy because the stress field of the individual dislocations no longer cancel each other so effectively. Such a boundary is most likely to be formed during plastic deformation. The long-range stresses can be removed by the boundary combining with a second wall of dislocations with Burgers vectors normal to the original ones as shown in the stable tilt boundary in Fig. 9.6. Similar arguments can be used to show that an infinite wall of parallel screw dislocations always has a long-range stress field, which is removed by the introduction of a second set of screw dislocations to form a cross grid as in Fig. 9.7.

As in the case of a single dislocation, work will be done when a dislocation is introduced into the stress field of a dislocation wall. The interaction force can be determined from equations such as (4.35) and (4.37) by using the appropriate stress summations. The problem may be complicated by the

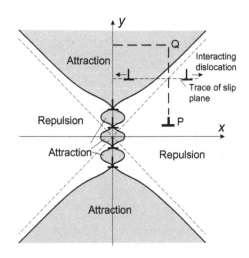

FIGURE 9.13

Dislocation wall containing four edge dislocations of the same sign. Similar dislocations of the same sign in parallel glide planes are attracted or repelled depending on their position.

movement of the boundary dislocations during the interaction. Some qualitative features are worth noting and, for simplicity, the intersection of an edge dislocation with a wall of edge dislocations of the same sign is considered, as depicted in Fig. 9.13. Near the wall, the dislocation experiences short-range forces similar to those for the infinite boundary (Fig. 9.12). In the attractive regions, this may result in a small displacement of an adjacent wall dislocation, so that the two 'share' one wall position. Alternatively, if the temperature is high enough, the wall dislocations may climb to accommodate the extra dislocation within the wall. The regions of attraction and repulsion are shown in Fig. 9.13. It is seen that at distances $\gtrsim D$ from the wall, attraction only occurs in the shaded regions delineated approximately by the lines $y = \pm x$. In these regions, the stress fields of the dislocations in the wall tend to reinforce each other rather than cancel: i.e. the dislocation in the matrix in Fig. 9.13 effectively experiences at large distances a force due to a single dislocation at the origin with Burgers vector four times that of the individual wall dislocations.

Figure 9.13 illustrates also a probable way in which a low-angle boundary develops during the *recovery process*. Dislocations in the unshaded regions will tend to be repelled by the boundary, but can climb by vacancy diffusion processes. If a dislocation at P climbs into the shaded region Q it will then experience an attractive force tending to align the dislocation in the low-energy configuration at the top of the existing wall. The path taken is indicated by the dotted line.

9.5 STRAIN ENERGY OF DISLOCATION ARRAYS

The energy per unit area of a low-angle boundary is found by multiplying the energy per unit length per dislocation by the number of dislocations per unit area, which is $D^{-1} = \theta/b$ for an array of parallel dislocations. The result, first published by Read and Shockley (see Read, 1953), is

$$E = E_0\theta(A - \ln \theta) \tag{9.15}$$

where constant E_0 is a function of the elastic properties of the material and A is a constant which depends on the core energy of an individual dislocation. An approximate derivation of this remarkably simple formula for the symmetrical tilt boundary of edge dislocations is as follows.

Consider the work done in creating one dislocation of the complete wall of identical dislocations. By analogy with the derivation of equation (4.22) in section 4.4, the contribution to the strain energy per unit length for the dislocation at the origin $y = 0$, for example, is

$$E(\text{per dislocation}) = \frac{1}{2}b \int_{r_0}^{\infty} \sigma_{xy}\, dx \qquad (9.16)$$

where σ_{xy} is given by equation (9.13) evaluated on $y = 0$, and r_0 is the dislocation core radius. It was shown in the preceding section that σ_{xy} for small x is approximately the stress due to a single dislocation at $y = 0$, i.e. $Gb/2\pi$ $(1 - \nu)x$, and that for large x it has the approximate exponential form of equation (9.14). The radius of the transition between these two expressions is proportional to D and is approximately $x = D/2\pi$. The contributions the two parts make to the energy are readily found from equation (9.16) to be

$$\frac{1}{2}b \int_{r_0}^{D/2\pi} \sigma_{xy}\, dx \simeq E_0 b \ln(D/2\pi r_0)$$
$$\qquad\qquad\qquad\qquad\qquad (9.17)$$
$$\frac{1}{2}b \int_{D/2\pi}^{\infty} \sigma_{xy}\, dx = E_0 b 2/e$$

where $E_0 = Gb/4\pi(1 - \nu)$. Multiplying the sum of these two terms by θ/b and replacing D by b/θ gives the Read–Shockley formula (equation (9.15)) with $A = \ln(b/2\pi r_0) + 2/e$. The term $2/e$ in A is replaced by 1 if the more accurate form of σ_{xy} is used.

This derivation emphasizes the fact that the $\ln(\theta)$ term arises from the *elastic* energy per dislocation. It *decreases* as θ *increases* because the stress fields of the dislocations overlap and cancel more fully as D decreases. If the non-linear dislocation core energy is added to equation (9.15), parameter A is modified in value but remains independent of θ. The Read–Shockley formula shows that it is energetically favorable for two low-angle boundaries with misorientation angles θ_1 and θ_2 to combine to form a single boundary with angle $(\theta_1 + \theta_2)$. Furthermore, it predicts that E has a maximum at $\theta = \exp(A - 1)$, as can be verified by differentiation of equation (9.15).

Experimental verification of equation (9.15) was one of the early successes of dislocation theory. The relative energy of a dislocation boundary was measured using the simple principle illustrated in Fig. 9.14. This shows three boundaries looking along their common axis. If E_1, E_2 and E_3 are the energies of the grain boundaries per unit area, each boundary will have an

FIGURE 9.14

Three grain boundaries that meet along a line (normal to the figure). Each boundary has an energy, E, per unit area which is a function of its misorientation angle, θ, and its crystallographic orientation.

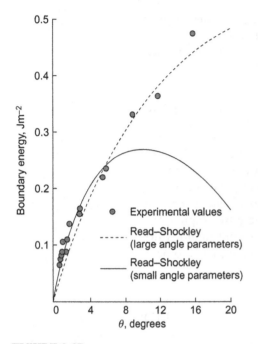

FIGURE 9.15

Relation between boundary energy and misorientation angle θ for [001] tilt boundaries in copper. Theoretical fits of the Read–Shockley equation using either high-angle or low-angle data are shown. *(After Gjostein and Rhines, Acta Metall. 7, 319 (1959).)*

effective surface tension equal to its energy. The situation will approximate, under equilibrium conditions, to a triangle of forces, and for the equilibrium of these forces acting at a point

$$\frac{E_1}{\sin \psi_1} = \frac{E_2}{\sin \psi_2} = \frac{E_3}{\sin \psi_3}$$

By measuring ψ_1 ψ_2 and ψ_3 and the misorientation across the boundaries it is possible to determine the relative energies.

Early measurements of E as a function of θ showed good agreement with the Read–Shockley formula, even for angles as large as $\sim 40°$. This was somewhat surprising, because for large-angle boundaries ($\theta \gtrsim 5-10°$), the dislocation cores overlap to such an extent that the derivation of equation (9.15) based on linear-elastic energy considerations ceases to be valid. Subsequent research on the energy of low-angle boundaries found that equation (9.15) does indeed give a good fit to the data for small θ, and that the apparent fit for high angles is fortuitous: it is only obtained by changing the constants E_0 and A from the values fitted at low angles. This is demonstrated by the data for tilt boundaries in copper in Fig. 9.15.

9.6 DISLOCATIONS AND STEPS IN INTERFACES

Admissible Defects

Line defects in interfaces between crystals can play an important role in transformations involving the growth of one crystal at the expense of another, e.g. twinning, martensitic transformations and precipitation. In the simplest picture, the defect is akin to a step and its motion over the interface is the mechanism that transforms atoms of one crystal to sites of the other. The word 'step' is misleading, however, because line defects in interfaces have specific character, which may or may not be step-like, and need to be defined with care. A brief description of this topic is presented here and more rigorous treatments are referenced at the end of the chapter.

The topological properties, e.g. the Burgers vector **b** and step height h, of interfacial defects that can arise when two crystals are joined to form a bicrystal are determined by the symmetry and orientation of the crystals. Specifically, they are defined by the crystal symmetries that are broken.

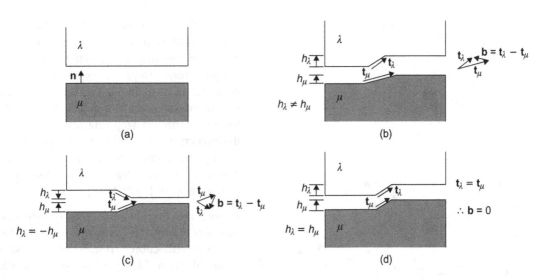

FIGURE 9.16
Illustration of the formation of a bicrystal by joining surfaces of crystals λ and μ. (a) Reference space has no interfacial defects. (b) Interfacial dislocation with step character and (c) one without. (d) Interface step with no dislocation character. The interface structures are identical on either side of the defect in (b)−(d).

An *interfacial dislocation* arises when either the translation or point group symmetry operations are broken. Consider the two crystals labeled μ and λ (black and white) in Fig. 9.16. If a bicrystal is created by bringing the upper planar surface of μ, with outward normal vector $\mathbf{n}$, into contact with the lower surface of λ, as in Fig. 9.16(a), a defect-free interface can result. If a surface contains a step and the surface structure is identical on either side of it, the step must be related to the crystal symmetry: in the simplest case, it is characterized by a translation vector $\mathbf{t}$ of the lattice, as indicated by the steps on the surfaces drawn in Figs 9.16(b)−(d). The height of such a step is $h = \mathbf{n} \cdot \mathbf{t}$ and can be signed positive or negative. For the situation in Fig. 9.16(b), the surfaces have steps that are neither parallel nor of the same height $(\mathbf{t}_\lambda \neq \mathbf{t}_\mu, h_\lambda \neq h_\mu)$, so that when they are brought together to form a bicrystal with no atoms removed and no spaces left along the stepped interface, the material in the vicinity of the resulting overlap step will have to be distorted to fit, as in the 'cutting and rebonding' process for crystal dislocations presented in section 3.2 Thus, a dislocation with Burgers vector

$$\mathbf{b} = \mathbf{t}_\lambda - \mathbf{t}_\mu \qquad (9.18)$$

is created, separating identical interfaces on either side. (Note that the sign convention used for $\mathbf{b}$ assumes that the positive line sense of the defect is

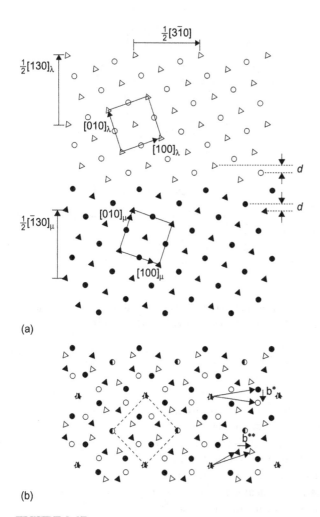

(a)

(b)

FIGURE 9.17

(a) Atom positions in two adjacent (001) planes in two face-centered cubic crystals separated by a symmetric {130}/⟨001⟩ 37°-tilt boundary. The unit cells are shown in outline and the atoms in different planes are plotted as circles and triangles. (b) Two lattices superimposed to show coincident sites, marked as half-filled symbols. The unit cell of this CSL with $\Sigma = 5$ is indicated by the dashed lines.

out of the paper in Fig. 9.16.) Interfacial defects with non-zero **b** and step height h have been termed *disconnections*. If the two surfaces have steps of equal height but opposite sign, as in Fig. 9.16(c), the resultant interface contains a defect that is a dislocation with no step character. Finally, if the two crystals have symmetry in common, i.e. $\mathbf{t}_\lambda = \mathbf{t}_\mu$, then it is possible to create a defect as a *pure step* without dislocation character, as in Fig. 9.16(d).

To illustrate these features in a specific case, consider a bicrystal formed by joining two face-centered cubic crystals of the same metal along a {130} plane of each after one has been rotated with respect to the other by an angle of $2\tan^{-1}(1/3) = 36.9°$ about a common ⟨001⟩ axis. The [001] projection of atoms that lie in two adjacent (001) planes is shown in Fig. 9.17(a). (The misorientation angle of this symmetric tilt boundary is too large for the interface structure to be described as an array of distinct dislocations, as in sections 2.3 and 9.2). Now imagine the two lattices to interpenetrate and bring a lattice point of each into coincidence by translation. This creates the *dichromatic pattern* of Fig. 9.17(b). It can be seen that for this particular misorientation of the crystals many sites (one in five) are in coincidence, shown by half-filled symbols. When such a situation occurs a *coincident-site lattice* (CSL) exists. The reciprocal of the ratio of CSL sites to lattice sites of one of the crystals is denoted by Σ, which is 5 in this case. CSLs arise most commonly for rotations about low-index directions in cubic crystals, but cannot be created so readily for lower symmetry systems or when crystals of different structure are joined, and so their general significance for models of interfaces should not be over-emphasized. Nevertheless, the example of Fig. 9.17 serves a useful purpose for illustrating the interfacial defects discussed above.

Since some of the translation vectors of each lattice are coincident – these are the vectors of the CSL – it is possible to form a pure step, as in Fig. 9.16 (d). This is demonstrated for the (130) interface in Fig. 9.18(a), where $\mathbf{t}_\lambda$ and $\mathbf{t}_\mu$ are $\frac{1}{2}[130]_\lambda$ and $\frac{1}{2}[\bar{1}30]_\mu$, respectively. Both h_λ and h_μ equal $10d$, where d is the interplanar spacing of the {130} planes drawn in Fig. 9.17(a). Although the indices of $\mathbf{t}_\lambda$ and $\mathbf{t}_\mu$ are different when expressed in the coordinate frame of their own lattice, they are identical when referred to a common frame, and so $\mathbf{b} = \mathbf{t}_\lambda - \mathbf{t}_\mu$ is zero. The defect is a step of height $10d$.

An interfacial dislocation with no step character, as in Fig. 9.16(c), can be formed by joining the λ and μ surfaces shown in Fig. 9.18(b). Here, $\mathbf{t}_\lambda$ is $\frac{1}{2}[2\bar{1}\bar{1}]_\lambda$ with height $h_\lambda = -d$ and $\mathbf{t}_\mu$ is $\frac{1}{2}[211]_\mu$ with $h_\mu = d$. The resultant defect has zero step height since $h_\lambda = -h_\mu$ but $\mathbf{b} = \mathbf{t}_\lambda - \mathbf{t}_\mu$ is not zero. It is a vector of the dichromatic pattern and is denoted as $\mathbf{b}^*$ in Fig. 9.17(b). In the coordinate frame of the white crystal, it is $\mathbf{b} = \frac{1}{10}[\bar{1}30]_\lambda$. An interfacial dislocation with step character, i.e. a disconnection as in Fig. 9.16(b), where the step has a small height can be formed as depicted in Fig. 9.18(c). In this case, $\mathbf{t}_\lambda$ is $[100]_\lambda$ with $h_\lambda = 2d$ and $\mathbf{t}_\mu$ is $\frac{1}{2}[110]_\mu$ with $h_\mu = 2d$. The resultant defect is a disconnection with step height $2d$ and $\mathbf{b} = \frac{1}{10}[3\bar{1}0]_\lambda$, which is denoted as $\mathbf{b}^{**}$ in Fig. 9.17(b). The dashed lines in Figs 9.18(b) and (c) indicate the extra half-planes and distortions associated with these edge dislocations.

It has been noted that both $\mathbf{b}^*$ and $\mathbf{b}^{**}$ are vectors from black to white sites of the dichromatic pattern. This is a general condition for the Burgers vector of dislocations that can exist in an interface of a bicrystal where translation symmetry is broken and the interface structure is identical on either side of the defect. If an interface can have more than one stable form, e.g. by rigid-body translation of one crystal with respect to the other, then an interfacial dislocation could separate structures that are not identical and its Burgers vector would not be a black-to-white vector of the dichromatic pattern. As noted above, CSLs arise only in special cases and are not a requirement for the approach to defect characterization presented here. The twin interface in a hexagonal-close-packed metal described in section 9.7 is an example where the reference space is not a three-dimensional lattice.

Identification of Interfacial Dislocations

The descriptions of the preceding section provide a means of defining the topological features of line defects that can arise in interfaces. Suppose, however, that a defect is observed in an interface by, say, high resolution electron microscopy: how can it be characterized? As with the Burgers circuit construction for a crystal dislocation (section 1.4), a circuit map can be used. A closed circuit around a feature to be identified is first made between

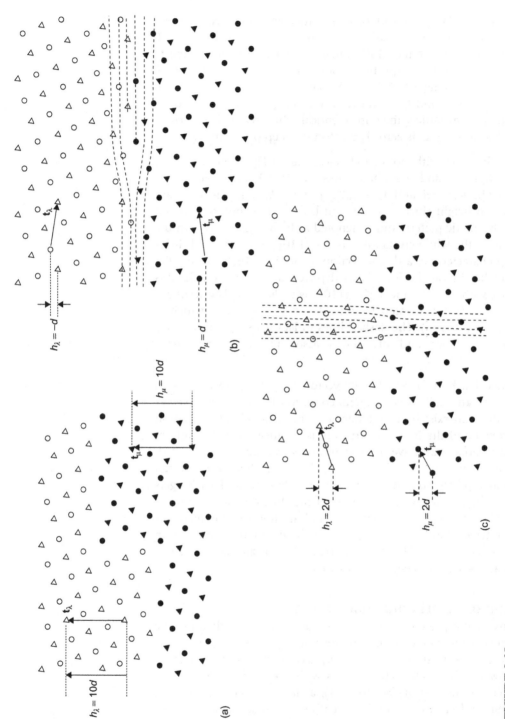

FIGURE 9.18

Some interfacial defects in a symmetric {130}/⟨001⟩ 37°-tilt boundary between face-centered cubic crystals: (a) pure step, (b) edge dislocation with no step and (c) edge dislocation with a step. Lattice vectors $\mathbf{t}_\lambda$ and $\mathbf{t}_\mu$ define the defect and its Burgers vector $\mathbf{b} = (\mathbf{t}_\lambda - \mathbf{t}_\mu)$. The horizontal and vertical dashed lines in (b) and (c) indicate the distortion associated with the dislocations.

crystallographically-equivalent white sites and then black sites, crossing the interface at two places. It is then mapped on to a reference space, but unlike the Burgers circuit for a crystal dislocation, there are two reference crystals, black and white. The presence of a closure failure ('finish-to-start') in this reference space identifies the Burgers vector content enclosed by the circuit in the real bicrystal.

Consider the disconnection line defect of Fig. 9.18(c). It is reproduced in Fig. 9.19(a) and, assuming that the defect has a positive sense out of the paper, a possible right-handed circuit starting and finishing at S is drawn on the figure. The steps XY and ZS cross the interface in equivalent ways and are not part of either crystal. The reference spaces taken from Fig. 9.18 are plotted in Fig. 9.19(b) and the circuit in (a) is mapped onto them. The presence of the closure failure confirms that the defect does indeed have dislocation character with $\mathbf{b} = \mathbf{FS}$, which is the same as $\mathbf{b}^{**}$ in Fig. 9.17(b).

Dislocations in Epitaxial Interfaces

The structure and defect content of the interface between a substrate and material grown on it by epitaxy is of considerable importance, particularly in the technology of solid-state electronic devices, and the methods of the previous two sections are useful for describing them. Consider for simplicity two cubic crystals λ and μ with their unit cells in parallel orientations, but let their lattice parameters a_λ and a_μ be different. If, for instance, $8a_\lambda = 7a_\mu$, as depicted in Fig. 9.20(a), the (001) interface is periodic because the arrangement shown repeats periodically along the interface. The admissible interfacial defects arising from the broken translation symmetry have $\mathbf{b} = (\mathbf{t}_\lambda - \mathbf{t}_\mu)$. The shortest lattice vectors are $\mathbf{t}_\lambda = [100]_\lambda$ and $\mathbf{t}_\mu = [100]_\mu$, and so $\mathbf{b}_{min} = \frac{1}{7}[\overline{1}00]_\lambda$ with $h = 0$.

Now consider deforming the two crystals until they are identical with $a_\lambda = a_\mu$, e.g. expand the λ crystal until the bicrystal in Fig. 9.20(b) can be formed. This is a *commensurate* interface, sometimes referred to as *coherent*. In dislocation terms, we can construct a closed circuit $SABCDE$ as in Fig. 9.20(b) and map it on to the λ and μ reference frames to test for a closure failure. This is shown in Fig. 9.20(c), and demonstrates that the circuit

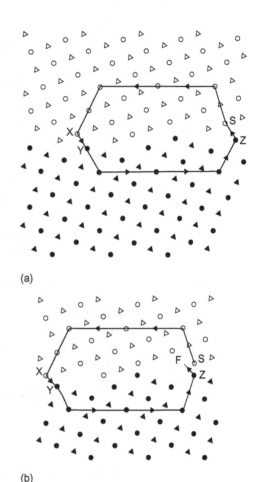

(a)

(b)

FIGURE 9.19
(a) Closed circuit drawn around the interfacial defect of Fig. 9.18(c). (b) Circuit of (a) mapped onto the reference space of the two crystals, showing that the defect has a Burgers vector **FS** ($=$ $\mathbf{b}^{**}$ in Fig. 9.17(b)).

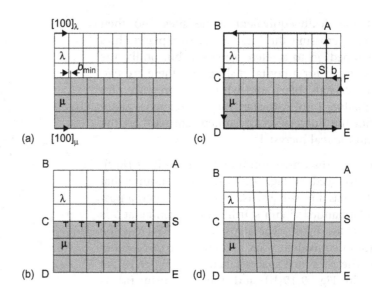

FIGURE 9.20

Illustration of the dislocation content of epitaxial interfaces: (a) reference λ and μ crystals; (b) crystal λ strained to produce a coherent interface which, with respect to (a) and as demonstrated in (c), contains the distribution of dislocations indicated by the usual symbols with $\mathbf{b} = \mathbf{b}_{min}$; (d) interface containing a 'misfit dislocation', but the dislocation content is actually zero. *(After Pond and Hirth, Fig. 2.6 of 'Defects at surfaces and interfaces', Solid State Physics, Vol. 47, p. 287 (1994)).*

in (b) does indeed enclose one or more dislocations with a total Burgers vector $\mathbf{b} = FS = [\bar{1}00]_\lambda$. The uniformity of the interface structure in Fig. 9.20(b) implies that this dislocation content may be considered to be distributed along the interface in the form of seven dislocations with $b = \frac{1}{7}[\bar{1}00]_\lambda = \mathbf{b}_{min}$. Their presence is indicated schematically in Fig. 9.20(b).

When a strained layer grows with a coherent interface, as in Fig. 9.20(b), the elastic strain energy increases in proportion to the volume, but can be relieved by the introduction of crystal dislocations (with $\mathbf{b} = \mathbf{t}_\lambda$ or $\mathbf{t}_\mu$) into the interface. They are known as *misfit dislocations* and enable the lattice parameter to return to the stress-free, equilibrium value. This is drawn schematically in Fig. 9.20(d) and an example obtained by high resolution electron microscopy for the interface between gallium arsenide and silicon is presented in Fig. 9.21. Note, however, that if the closed circuit *SABCDE* of Fig. 9.20(d) is mapped onto the reference space of Fig. 9.20(c), no closure failure results. In other words, the structure of Fig. 9.20(d) has zero dislocation content because the long-range stress field of the misfit dislocation is

FIGURE 9.21
(a) High-resolution electron microscope image revealing the atomic structure of the interface between gallium arsenide and silicon which contains misfit dislocations (arrowed). (b) Tilted and transformed image showing the presence of the misfit dislocations more clearly. *(Courtesy R. Beanland and A. F. Calder.)*

annulled by the uniform distribution of dislocations of opposite sign (Fig. 9.20(b)), resulting in a total **b** of zero.

9.7 MOVEMENT OF BOUNDARIES

For the movement of a boundary containing interfacial dislocations, several conditions must be satisfied. Firstly, if slip is involved the dislocation must be free to glide on its slip plane, i.e. the plane that contains the line and its Burgers vector. Secondly, matter must be conserved if a diffusional flux of atoms is required for motion of interfacial defects. Thirdly, the thermodynamic condition that movement results in either a reduction in the energy of the boundary or, in the case of movement induced by an externally applied stress, that the stress does work when the boundary moves. Fourthly, the driving force due, for example, to the externally applied stress or the excess vacancy concentration must be sufficient to produce dislocation movement. Some aspects of the stress required to move dislocations are considered in the next chapter.

Arrays of Crystal Dislocations

Consider first the low-angle boundaries formed by dislocations whose Burgers vectors are lattice vectors (sections 9.2 and 9.3). The condition for slip is particularly restrictive on the movement of such boundaries and can

(a)

(b)

FIGURE 9.22

Stress-induced movement of a symmetrical pure tilt boundary.

be illustrated with reference to the motion of low-angle tilt boundaries entirely by slip: the dislocations must have parallel glide planes. Consider the symmetrical tilt boundary illustrated in Fig. 9.22(a). A shear stress can be applied to the boundary by adding a weight to one end of the crystal as illustrated. For a shear stress τ, acting on the slip planes in the slip direction, the force on every dislocation will be $b\tau$ per unit length, and since there are θ/b dislocation lines per unit boundary height, the force per unit area on the boundary is

$$F = \theta\tau \tag{9.19}$$

If this is sufficient to move the dislocations the boundary will move to the left as illustrated in Fig. 9.22(b). Since every dislocation remains in the same position relative to the others in the boundary the geometry of the boundary is conserved. The work done by τ is $\tau\theta$ per unit volume swept out by the boundary. The movement of such a boundary was observed directly by Washburn and Parker (1952) in single crystals of zinc. Tilt boundaries were introduced by bending and annealing to produce polygonization as in Fig. 9.3. The boundary was moved backwards and forwards by reversing the direction of the applied stress.

Now consider the tilt boundary illustrated in Fig. 9.6, in which the Burgers vectors of the component edge dislocations are at right-angles. There are three possibilities for the movement of such a boundary under an applied stress. (a) The dislocations move by glide and remain in the same plane parallel to the original boundary. Since the slip planes are orthogonal the resolved shear stresses on the two systems are identical for a particular uniaxial stress, but induce the two sets of dislocations to move in opposite directions, as illustrated in Fig. 9.23. Thus, if all the dislocations were to move together, the net work performed would be zero and there would be no tendency for this to occur. (b) The dislocations move by glide but in opposite directions (Fig. 9.23). If both sets of dislocations move in such a way that work is done, the boundary will tend to split in the way illustrated, whereas the forces between dislocations will tend to keep the two boundaries together in the minimum energy configuration. Extra energy will be required and the process will be difficult. (c) The boundary moves uniformly as a whole, such that the dislocation arrangement is conserved. This can occur only by a combination of glide and climb. The stress field indicated on Fig. 9.23 shows that the set of dislocations (1) is under a compressive stress and will tend to climb up and to the right by the addition of vacancies. The set of dislocations (2) is under a tensile stress and will tend to climb down and to the right by emission of vacancies. Thus by a combination of climb and glide the boundary can move as a whole and the vacancies created at

FIGURE 9.23
Stress-induced movement of a tilt boundary containing edge dislocations with mutually perpendicular glide planes. *(After Read (1953), Dislocations in Crystals, McGraw-Hill.)*

one set of dislocations are absorbed by the other. Only short-range diffusion will be required and the process can occur at a high enough temperature.

Apart from the tilt boundary discussed above, and illustrated in Fig. 9.22, the only low-angle boundary that can move entirely by glide is a cross grid of screw dislocations and in this case it is essential that the junctions do not dissociate (see Fig. 7.21). In general, for stress induced boundary movement some diffusion is required, otherwise the boundary tends to break up, such as when a polygonized structure is deformed at low temperature.

Glide of Interfacial Defects

The discussion above is concerned with crystal dislocations. Now consider boundaries containing more general interfacial defects (section 9.6). The glide plane of a dislocation whose Burgers vector lies in the interface is the plane of the interface itself. A disconnection (Fig. 9.16(b)) with $h_\lambda = h_\mu$, as in Fig. 9.18(c), meets this condition and can glide if the resolved shear stress is high enough. Glide of the step allows one crystal to grow at the expense of the other: for example, glide of the defect in Fig. 9.18(c) to the right would allow black atoms to transfer to sites in the white crystal as the step passes. The simplest, yet very important, example of this occurs in *deformation twinning*.

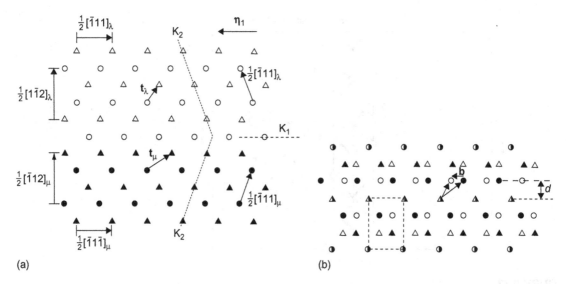

FIGURE 9.24

(a) Projection of the atomic positions in two adjacent (110) planes in a bicrystal formed by ⟨111⟩{112} twinning in a body-centered cubic metal. The twin habit plane is K_1, the direction of the twinning shear is η_1 and [110] is out of the paper. (b) Dichromatic pattern associated with (a). $\mathbf{b} = (\mathbf{t}_\lambda - \mathbf{t}_\mu)$ is the Burgers vector of a twinning dislocation formed by joining steps $\mathbf{t}_\lambda$ and $\mathbf{t}_\mu$ on the two crystals, as indicated.

Deformation twinning on the ⟨111⟩{11$\bar{2}$} system is common in the body-centered cubic metals, as noted in section 6.3. The direction and magnitude of the twinning shear is consistent with the glide of dislocations with $\mathbf{b} = \frac{1}{6}\langle 111\rangle$ on every successive {11$\bar{2}$} atomic plane. This can be understood in terms of the description of admissible interfacial defects given in section 9.6. The defect-free interface is plotted in [110] projection in Fig. 9.24(a). (Visualization of the atomic positions shown is assisted by comparison with Fig. 1.6.) The twin habit plane is K_1 and the twinning shear in the direction η_1 reorientates the complementary twinning plane K_2 as indicated. The dichromatic pattern for the bicrystal is plotted in Fig. 9.24(b), which is seen to be a CSL with $\Sigma = 3$. The interfacial defect with a step up (from left to right) and height d, the spacing of the {112} planes, and the shortest Burgers vector lying parallel to the interface, is created by joining the steps defined by $\mathbf{t}_\lambda = \frac{1}{2}[\bar{1}\bar{1}1]_\lambda$ and $\mathbf{t}_\mu = \frac{1}{2}[010]_\mu$ as indicated. This disconnection has $\mathbf{b} = \frac{1}{6}[1\bar{1}\bar{1}]_\lambda$ and is a twinning dislocation, as described in section 6.3, and, with the convention that positive line sense is out of the paper (section 9.6), is a negative edge dislocation. For a step of opposite sign, i.e. step down from left to right, the defect would be a positive dislocation.

Thus, the atomic displacements that give rise to the macroscopic twinning shear occur by the glide of these defects on successive planes, as illustrated

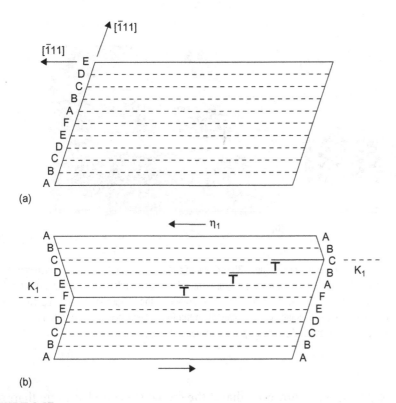

FIGURE 9.25

Schematic illustration of twinning in a body-centered cubic metal. The projection is the same as that in Fig. 9.24 and the dislocations are the same as the one defined there. (a) Untwinned crystal in [110] projection showing the stacking sequence $A, B, \ldots$ of the $(\overline{1}12)$ planes. (b) Twinning dislocations with $\mathbf{b} = \frac{1}{6}[1\overline{1}1]_{\mu}$ glide to the right on successive $(\overline{1}12)$ planes under the applied shear stress indicated to produce a twin-orientated region.

schematically in Fig. 9.25. An experimental observation of dislocations of this type is presented in Fig. 9.26. Figure 9.26(a) illustrates the cross-section shape of the small twin seen in the transmission electron microscope image in (b) and an explanation for the contrast from the individual twinning dislocations is sketched in (c).

Twinning is an important mode of deformation for many crystalline solids, particularly if the number of independent slip systems associated with the glide of crystal dislocations is restricted (section 10.9). In all cases, the $\mathbf{b}$ and h characteristics of the twinning dislocations can be analyzed using treatments similar to those presented here. (Further details can be found in references at the end of the chapter.) The glide motion of twinning dislocations can occur under low resolved shear stress in many materials and is therefore

FIGURE 9.26
Experimental observation of a small deformation twin in a molybdenum—35% rhenium alloy by transmission electron microscopy. (a) Illustration of the shape of the small twin shown in (b); the dislocations are represented by dots. (b) Diffraction contrast produced by twin which lies at an angle of 20° to the plane of the thin foil. (c) Diagrammatic illustration of the diffraction contrast observed in (b). The K$_1$ plane is (112) and the dislocations are in screw orientations. Each change in the fringe sequence is due to a twinning dislocation. *(From Hull (1962)*, Proc. 5th Int. Conf. Electron Microscopy, p. B9, Academic Press.)

relatively easy. Note, however, that if the crystal structure has more than one atom per lattice site, then the simple shear associated with the passage of such dislocations may restore the lattice in the twinned orientation but not all the atoms. Shuffles of these atoms will be required. This is illustrated by the $\{10\bar{1}2\}$ twin in an hexagonal-close-packed metal, which has two atoms per lattice site. The structure of the twinning dislocation found by computer simulation (section 2.4) is shown in Fig. 9.27(a). The positive edge dislocation defined by $t_\lambda - t_\mu$ results in a step down (from left to right) of height $2d$, where d is the spacing of the $\{10\bar{1}2\}$ lattice planes.

This structure has been verified experimentally, as demonstrated by the high resolution transmission electron microscopy image in Fig. 9.27(b). The black dots indicate the match between the positions of atoms near the interface in this image and those in Fig. 9.27(a). It can be seen from Fig. 9.27(a) that the atoms in the two atomic planes labelled S that traverse the step have to shuffle as the dislocation glides along the boundary, because atoms such as 1 and 2 are closer than 2 and 3 on the left whereas 2 and 3 are closer than 1 and 2 on the right. The shuffles are short and easily achieved in this particular case, and so the core of the dislocation spreads along the interface. Computer simulation shows that this twinning dislocation moves easily. For boundaries where complex shuffles are necessary, the glide of twinning

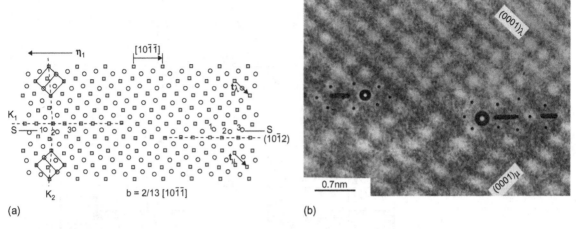

FIGURE 9.27

(a) Atomic structure obtained by computer simulation of the structure of a twinning dislocation in a $(10\bar{1}2)$ twin boundary in titanium, an hexagonal-close-packed metal. Unit cells are shown in outline and the position of the boundary is indicated by a dashed line. The twinning dislocation, defined by the lattice vectors $\mathbf{t}_\lambda$ and $\mathbf{t}_\mu$, has a very small Burgers vector, but requires shuffling of atoms in the layers labelled S. (*After Bacon and Serra (1994), Twinning in Advanced Materials, eds. M. H. Yoo and M. Wuttig, p. 83. The Minerals, Metals and Materials Society (TMS).*) (b) Experimental HRTEM image of a boundary in titanium containing a twinning dislocation. The dashed lines show the location of the interface and the dots indicate the positions of some atoms near the interface. (*From Braisaz, Nouet, Serra, Komninou, Kehagias and Karakostas, Phil Mag. Letters **74**, 331 (1996), with permission from Taylor and Francis Ltd (http://www.tandf.co.uk/journals).*)

dislocations can require relatively high stress and the assistance of elevated temperature.

Diffusion-assisted Motion of Interfacial Defects

The slip plane of the interfacial dislocations described in the preceding section is the plane of the interface itself and so they glide under the action of a resolved shear stress. If the dislocation has a Burgers vector with a component perpendicular to the interface, however, it can move along the boundary by climb, irrespective of the nature of the two crystals, and so requires the diffusion of atoms to or from it. Furthermore, when either the atomic density or the chemical composition of the two crystals is different, i.e. the interface is an *interphase boundary*, a defect with step character can only move along the interface if atomic transport can occur. These two statements indicate that analysis of the atomic flux required for motion of interfacial defects has to treat both the dislocation and step components of the defect. This is demonstrated by the following simple example.

For the step drawn in Fig. 9.28, $\mathbf{t}_\lambda$ and $\mathbf{t}_\mu$ are parallel but of different length, and so $h_\lambda \neq h_\mu$ and the dislocation has a Burgers vector component b_n

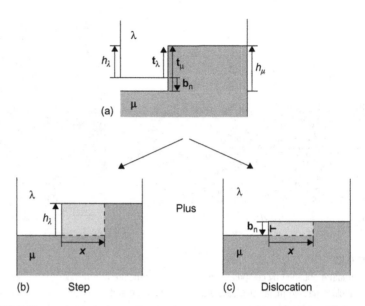

FIGURE 9.28

Step and dislocation parts of the interfacial defect in the bicrystal in (a) are shown in (b) and (c). As the defect moves to the right along the interface the volume of the λ crystal increases at the expense of the μ crystal because of the step component and the volume of the μ crystal decreases due to the climb of the dislocation component.

perpendicular to the interface. Consider the step part first (Fig. 9.28(b)). If it moves a distance x to the right, the volume change of the λ crystal is

$$V = h_\lambda x \tag{9.20}$$

per unit length of step. Let λ have atoms of species A, B, C, ... with X_λ^A, X_λ^B, X_λ^C, ... atoms per unit volume, and similarly for crystal μ. The change in the number of A atoms in the volume swept by the step is then

$$\Delta N^A = h_\lambda x (X_\lambda^A - X_\mu^A) \tag{9.21}$$

For the step to move with a velocity v, ΔN^A atoms will have to arrive or leave in the time x/v, and so the diffusional flux (atoms per second per unit step length) will have to be

$$J^A = h_\lambda v (X_\lambda^A - X_\mu^A) \tag{9.22}$$

Now consider the dislocation part of the defect (Fig. 9.28(c)). As shown by the description of climb in section 3.6, the volume change of the μ crystal is

$$V = b_n x \tag{9.23}$$

per unit length of dislocation. The number of A atoms involved is

$$\Delta N^A = b_n x X_\mu^A \tag{9.24}$$

and the diffusive flux to the dislocation is

$$J^A = b_n \nu X_\mu^A \tag{9.25}$$

Thus, from equations (9.22) and (9.25) the net flux of A atoms to the complete defect is

$$J^A = \nu[h_\lambda(X_\lambda^A - X_\mu^A) + b_n X_\mu^A] \tag{9.26}$$

and similarly for B, C, ... (Note that V, ΔN^A and J^A are negative ($b_n < 0$) in equations (9.23–9.25) because the dislocation climb shown removes μ crystal. It does not create λ crystal: that is done by motion of the step. Care must be exercised in choosing an unambiguous sign convention for the step height and b_n for use in the equations above.)

For interfacial defect motion along *grain boundaries*, of which the twin interface is a special case, the two crystals have identical chemical composition and atomic density and so $X_\lambda^A = X_\mu^A$. Equation (9.26) shows that motion is controlled solely by the normal component b_n of the Burgers vector, irrespective of whether a step exists or not. For motion along an *interphase boundary*, $(X_\lambda^A - X_\mu^A)$ is not zero in general because even if the chemical composition of the two phases is the same, a difference in crystal structure usually results in a difference in the number of atoms per unit volume. Treatments based on equation (9.26) for some specific material systems are presented in references at the end of the chapter.

Martensitic Transformations

These transformations are used to develop desirable properties in many materials, such as hardenability in steels and shape-memory effects in alloys. They occur during cooling to a phase that is more stable than the parent phase at low temperature, but, unlike precipitation (section 10.6), the transformation is not one of nucleation followed by diffusion-controlled growth. The displacement of atoms with respect to their neighbors is small ($\leq$ interatomic spacing) and motion of the interface between parent and product phases occurs without diffusion, i.e. the transformation is *displacive*. In this respect the transformation is similar to twinning, except that the crystal structure and atomic density of the product are different from those of the parent. The mechanism of displacive transformations is the motion of disconnections (interfacial defects with both dislocation and step character – see section 9.6). The shape change associated with the martensitic transformation

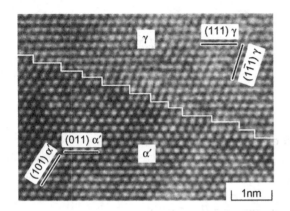

FIGURE 9.29

High resolution micrograph of interface between martensite (α') and austenite (γ) in Fe−20.2Ni−5.4 Mn. The {110}α'/{111}γ terraces and the disconnection lines are perpendicular to the paper. *(Reproduced from Moritani, Miyajima, Furuhara and Maki, Scripta Mater. 47, 193, 2002. Copyright (2002) with permission from Elsevier.)*

is not pure shear, however, and the restrictions on the orientation of the *habit plane* between the phases and the mechanisms that allow it to move are more severe than in twinning.

Recent studies by high-resolution transmission electron microscopy show that the interface consists of terraces, as illustrated by the example in Fig. 9.29. A *topological model* has been developed to identify the orientation of the terrace plane and the line defects present in it, subject to the condition that the semi-coherent interface can move conservatively. The general form of the interface is drawn in Fig. 9.30. It consists of two sets of parallel line defects forming an array bounding the terrace segments. Slip or twinning dislocations glide in the martensite crystal; they deform the lattice but leave it invariant. Disconnections form the other set. When the disconnection array moves laterally, the interface advances by *h*. The topological model predicts the spacing, Burgers vector and step height of these defects such that the interface is able to advance without diffusion. (See Further Reading for more detail.)

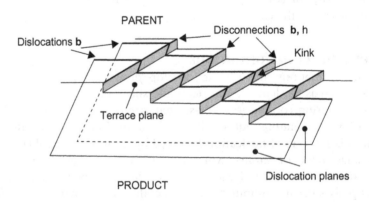

FIGURE 9.30

Schematic illustration of parent-martensite interface showing that the habit plane consists of terraces and arrays of parallel disconnections and parallel dislocations. *(After Pond, Ma, Chai and Hirth (2007), Dislocations in Solids, vol. 13, p. 225 (eds. F.R.N. Nabarro and J. P. Hirth), North-Holland). Copyright (2007) with permission from Elsevier.)*

9.8 DISLOCATION PILE-UPS

The dislocation pile-up is a completely different kind of dislocation array that forms during plastic deformation. Thus, consider a dislocation source which emits a series of dislocations all lying in the same slip plane (Fig. 9.31). Eventually the leading dislocation will meet a barrier such as a grain boundary or sessile dislocation configuration and further expansion of the loop is prevented. The dislocations then pile-up behind the leading dislocation, but, being of the same sign, do not combine. They interact elastically and their spacing, which depends on the applied shear stress and the type of dislocation, decreases towards the front of the pile-up.

FIGURE 9.31
Linear arrays of edge dislocations piled-up against barriers under an applied shear stress τ.

The stress τ_1 experienced by the leading dislocation of a pile-up can be deduced as follows. Suppose there are n dislocations. The leading dislocation experiences a forward force due to the applied stress τ and the other $(n-1)$ dislocations, and a backward force due to the internal stress τ_0 produced by the obstacle. If the leading dislocation moves forward by a small distance δx, so do the others, and the applied stress does work per unit length of dislocation equal to $nb\tau\delta x$. The increase in the interaction energy between the leading dislocation and τ_0 is $b\tau_0\delta x$. In equilibrium, these energies are equal and $\tau_1 = \tau_0$, so that

$$\tau_1 = n\tau \tag{9.27}$$

Thus, the stress at the head of the pile-up is magnified to n times the applied stress. The pile-up exerts a back-stress τ_b on the source, which can only continue to generate dislocations provided $(\tau - \tau_b)$ is greater than the critical stress for source operation. Eshelby et al. (1951) calculated the spatial distribution of dislocations in a pile-up and found that for a single-ended pile-up spread over the region $0 \le x \le L$, the number of dislocations in the pile-up is

$$n = \frac{L\tau}{A} \tag{9.28}$$

where A is Gb/π for screw dislocations and $Gb/\pi(1-\nu)$ for edges. The spacing of the first two dislocations is $0.92 L/n^2$. The shear stress outside the pile-up well away from the first and last dislocations is approximately the same as that produced by a single superdislocation of Burgers vector nb at the center of gravity of the pile-up ($x = 3L/4$).

Thus, unlike the Read–Shockley arrays studied in the first part of this chapter, dislocation pile-ups produce large, long-range stress. At grain boundaries, this can nucleate either yielding in the adjacent grains (section 10.9) or boundary cracks. It can also assist in the cross slip of screw dislocations held up at obstacles such as precipitates and dislocation locks. Dislocation pile-ups have been observed many times using transmission electron microscopy. Most pile-ups will be made up of dislocations with an edge component to the Burgers vector since screw dislocations can cross slip out of the slip plane. It should also be noted that certain crack configurations can be modeled in elasticity theory by pile-ups.

FURTHER READING

Cherkaoui M, Capolungo L: *Atomistic and continuum modeling of nanocrystalline materials: deformation mechanisms and scale transition*, 2009, Springer.

Christian JW: *The theory of transformations in metals and alloys (Part I and II)*, ed 3, 2002, Pergamon.

Christian JW: Deformation by moving interfaces, *Metall Trans* A13:509, 1982.

Christian JW, Crocker AG: Dislocations and lattice transformations. In Nabarro FRN, editor: *Dislocations in solids*, vol 3, 1980, North-Holland, p. 165.

Demkowicz MJ, Wang J, Hoagland RG: Interfaces between dissimilar crystalline solids. In Hirth JP, editor: *Dislocations in solids*, vol 14, 2008, North-Holland, p. 141.

Eshelby JD, Frank FC, Nabarro FRN: The equilibrium of linear arrays of dislocations, *Phil Mag* 42:351, 1951.

Frank FC: *Conf. on plastic deformation of crystalline solids*, 1950, Carnegie Inst. of Tech. and Office of Naval Research, p. 150.

Hirth JP, Lothe J: *Theory of dislocations*, 1982, Wiley.

Matthews JW: Misfit dislocations. In Nabarro FRN, editor: *Dislocations in solids*, vol 2, 1979, North-Holland, p. 461.

Pond RC, Hirth JP: Defects at surfaces and interfaces, *Solid state physics* vol 47:287, 1994.

Pond RC, Ma X, Chai YW, Hirth JP: Topological modelling of martensitic transformations. In Nabarro FRN, Hirth JP, editors: *Dislocations in solids*, vol 13, 2007, North-Holland, p. 225.

Niewczas M: Dislocations and twinning in face-centred cubic crystals. In Nabarro FRN, Hirth JP, editors: *Dislocations in solids*, vol 13, 2007, North-Holland, p. 263.

Read WT: *Dislocations in crystals*, New York, 1953, McGraw-Hill.

Read WT, Shockley W: Dislocation models of grain boundaries, In *Imperfections in nearly perfect crystals*, 1952, Wiley, p. 352.

Sutton AP, Balluffi RW: *Interfaces in crystalline materials*, 1995, Oxford University Press.

Washburn J, Parker ER: Kinking in zinc single-crystal tension specimens, *Trans Amer Inst Min (Metall) Engrs* 194:1076, 1952.

CHAPTER 10

Strength of Crystalline Solids

10.1 INTRODUCTION

In Chapter 1 it was shown that the theoretical shear strength of a perfect crystal is many orders of magnitude greater than the observed critical shear strength of real crystals which contain dislocations. In this chapter the factors affecting the strength of crystals are considered as an introduction to the way that an understanding of the properties of dislocations can be used to interpret the behavior of crystalline solids.

Apart from effects associated with diffusion at high temperature, plastic deformation occurs by the glide of dislocations and hence the critical shear stress for the onset of plastic deformation is the stress required to move dislocations. This is measured usually by a tensile test in which the specimen is elongated at a *constant rate* and the load on the specimen is measured simultaneously with the extension. Representations of typical stress–strain curves are shown in Fig. 10.1. Curves (a) and (b) are typical of many solids: (a) represents a *ductile* material which undergoes extensive plastic deformation before fracture at *F*, and (b) represents a *brittle* material which exhibits little plasticity. The dislocations in (b) are either too low in density or too immobile to allow the specimen strain to match the elongation imposed by the testing machine. The curves show a linear region *OE* in which the specimen deforms elastically, i.e. the stress is proportional to strain according to Hooke's Law, followed by *yielding* at *E* and subsequent *strain* (or *work*) *hardening* up to *F*. In the latter process, the *flow stress* required to maintain plastic flow increases with increasing strain. For most materials, the change from elastic to plastic behavior is not abrupt and the *yield stress* σ_y is not unique. This is because some non-linear *microplasticity* occurs in the pre-yield region *OE* due to limited dislocation motion. Then, as shown in Fig. 10.1(a), yielding is defined to occur when the plastic strain reaches a prescribed value, say 0.1 per cent: the corresponding *proof stress* is taken as the yield stress. Figure 10.1(c) is typical of many body-centered cubic polycrystalline metals

205

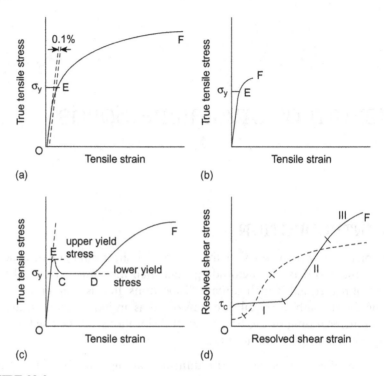

FIGURE 10.1

Typical stress–strain curves for (a), (b), (c) polycrystals and (d) single crystals.

which do not yield uniformly. An example for iron is shown in Fig. 10.17 (b). The curve can be divided into four regions, namely: (*OE*) elastic and pre-yield microplastic deformation, (*EC*) yield drop, (*CD*) yield propagation, and (*DF*) uniform hardening. The deformation between *E* and *D* is not homogeneous, for plastic flow occurs in only part of the specimen. This *Lüders band*, within which dislocations have rapidly multiplied, extends with increasing strain to occupy the entire length at *D*. In tests on single crystals, it is usual to resolve the stress and strain onto the plane and direction for which slip occurs first. The resulting stress–strain curve often has the form shown in Fig. 10.1(d). An actual example for copper is reproduced in Fig. 10.17(a). Above the *critical resolved shear stress* τ_c (section 3.1), the curve has three parts: stages I, II and III. The extent of these stages depends on the material and crystal orientation of the load axis, as indicated by the two curves. Polycrystalline specimens of the same metal (Fig. 10.1(a)) do not show stage I but deform in a manner equivalent to stages II and III.

It should be noted that the form of all the curves in Fig. 10.1 is dependent on test variables such as temperature and applied strain rate. Material

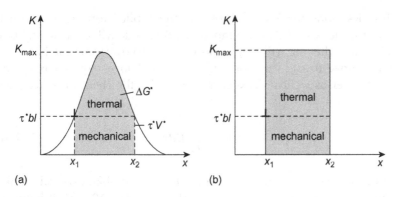

FIGURE 10.2
Profiles of resistance force K versus distance x for barriers opposing dislocation motion.

parameters such as crystal structure, alloy composition, dislocation arrangement and grain size also affect the yield and flow stresses. It is therefore possible to modify materials in order to improve their performance. The
understanding of these effects has been one of the successes of dislocation
theory and is the subject of this chapter. In section 10.2 the manner in
which the strength of barriers to dislocation motion affects the dependence
of the flow stress on temperature and strain rate is considered. The nature of
these barriers and the ways in which they are introduced are treated in the
following sections. The presentation is, in part, necessarily simplified, and
phenomena such as creep and fatigue are not touched upon at all. Reviews
and texts with comprehensive descriptions are given in Further Reading.

10.2 TEMPERATURE- AND STRAIN-RATE-DEPENDENCE OF THE FLOW STRESS

The energy that has to be provided for dislocations to overcome the barriers
they encounter during slip determines the dependence of the flow stress on
temperature and applied strain rate. If the energy barriers are sufficiently
small for thermal energy ($\sim kT$) to be significant, thermal vibrations of the
crystal atoms may assist dislocations to overcome obstacles at lower values
of applied stress than that required at 0 K. Under such conditions, an
increase in temperature, or a reduction in applied strain rate, will reduce the
flow stress. The principles involved can be demonstrated as follows.

Consider a dislocation gliding in the x direction under an applied resolved
shear stress τ^*, which produces a force $\tau^* b$ per unit length on the line.
Suppose the dislocation encounters obstacles, each of which produces a
resisting force K, as shown schematically in Fig. 10.2(a). Let the spacing of

the obstacles along the line be l, so that the applied forward force on the line per obstacle is τ^*bl. At the temperature 0 K, glide will cease if τ^*bl is less than K_{max}, and the line will stop at position x_1. To overcome the barrier, the line must move to x_2. The isothermal energy change required is the area under the K versus x curve between x_1 and x_2, i.e. the Helmholtz free energy change,

$$\Delta F^* = \int_{x_1}^{x_2} K dx \tag{10.1}$$

Part of this energy can be provided in the form of mechanical work done by the applied load, and is $\tau^*bl(x_2 - x_1)$, as shown in Fig. 10.2(a). It is customary to rewrite this mechanical contribution as τ^*V^*, where V^* is known as the *activation volume* for the process. The remainder of the energy required is labeled 'thermal' in Fig. 10.2(a). It is the *free energy of activation*:

$$\Delta G^* = \Delta F^* - \tau^*V^* \tag{10.2}$$

and is the Gibbs free energy change between the two states x_1 and x_2 at the same temperature and applied stress. The probability of energy ΔG^* occurring by thermal fluctuations at temperature T (such that $\Delta G^* \gg kT$) is given by the Boltzmann factor $\exp(-\Delta G^*/kT)$, so that if the dislocation is effectively vibrating at a frequency ν ($\lesssim$ atomic vibration frequency), it successfully overcomes barriers at a rate of $\nu \exp(-\Delta G^*/kT)$ per second. The dislocation velocity is therefore

$$\bar{v} = d\nu \exp(-\Delta G^*/kT) \tag{10.3}$$

where d is the distance moved for each obstacle overcome. From equations (3.13) and (10.3), the macroscopic plastic strain rate is

$$\dot{\varepsilon} = \rho_m A \exp(-\Delta G^*/kT) \tag{10.4}$$

where ρ_m is the mobile dislocation density and $A = bd\nu$. The stress-dependence of $\dot{\varepsilon}$ arises from the stress-dependence of ΔG^*. The resulting relation between flow stress and temperature can be derived simply for the following case.

Suppose that the obstacles form a regular array and that each one is a region of constant resisting force, as shown in Fig. 10.2(b). Then

$$\Delta G^* = \Delta F\left[1 - \frac{\tau^*(T)}{\tau^*(0)}\right] \tag{10.5}$$

where ΔF is the total area under the K versus x curve, i.e. the energy required to overcome the obstacle when $\tau^* = 0$, and $\tau^*(0)$ is the flow stress at 0 K, i.e. the stress required to overcome the obstacle when no thermal energy is available. Substituting equation (10.5) into (10.4) gives for the flow stress

$$\frac{\tau^*(T)}{\tau^*(0)} = \frac{kT}{\Delta F}\ln\left(\frac{\dot\varepsilon}{\rho_m A}\right) + 1 \qquad (10.6)$$

At temperatures above a certain value, say T_c, there will be sufficient thermal energy for the barriers to be overcome by thermal activation alone. Then $\tau^*(T_c) = 0$ and from equation (10.6)

$$T_c = \frac{-\Delta F}{k\,\ln(\dot\varepsilon/\rho_m A)} \qquad (10.7)$$

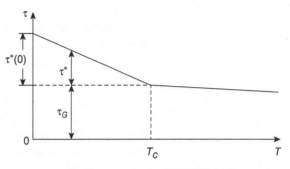

FIGURE 10.3
Variation of the flow stress with temperature for obstacles represented by Fig. 10.2(b).

Substituting this into equation (10.6) gives

$$\frac{\tau^*(T)}{\tau^*(0)} = \left(1 - \frac{T}{T_c}\right) \qquad (10.8)$$

Hence, as shown in Fig. 10.3, τ^* decreases from $\tau^*(0)$ to zero as T increases from 0 K to T_c.

Generally, the motion of a dislocation is opposed by short-range barriers, which can be overcome by thermal activation as described here, and long-range forces from, say, other dislocations. The latter are proportional to the shear modulus G and produce barriers too large for thermal activation to be significant. The flow stress then consists of *two* contributions. One is the *thermal component* τ^*: the other is the *athermal component* τ_G, which is almost independent of temperature apart from the small variation of G with temperature. Thus

$$\tau = \tau^* + \tau_G \qquad (10.9)$$

The result of combining equations (10.8) and (10.9) is shown in Fig. 10.3. The temperature-dependence of τ for $T < T_c$ depends on the size of $\tau^*(0)$, which is proportional to K_{max}, and thus on the nature of the obstacles. Most of the barriers encountered by dislocations do not have a simple square profile, as considered in Fig. 10.2(b), and do not form regular arrays. In more general situations, equation (10.5) is replaced by

$$\Delta G^* = \Delta F\left\{1 - \left[\frac{\tau^*(T)}{\tau^*(0)}\right]^p\right\}^q \qquad (10.10)$$

with $0 \leq p \leq 1$ and $1 \leq q \leq 2$. This generally results in a curved rather than straight-line relationship between τ^* and T. The energy ΔF lies between about 0.05 Gb^3 and 2 Gb^3. This encompasses the range from small barriers such as solute atoms and the Peierls stress up to large barriers such as strong precipitates.

The approach of this section illustrates the features which control the flow stress at low to moderate temperatures. In particular, an increase in temperature or a decrease in applied strain rate provides an increase in the probability of thermal activation and therefore results in a reduction in flow stress. It should also be noted, however, that when ΔG^* is very small ($\simeq kT$), the Boltzmann distribution is no longer valid. Dislocation velocity is then determined not by thermally-activated release from obstacles, but by lattice dynamical effects, as discussed in section 3.5.

10.3 THE PEIERLS STRESS AND LATTICE RESISTANCE

The applied resolved shear stress required to make a dislocation glide in an otherwise perfect crystal without the assistance of thermal activation, i.e. at $T = 0$ K, is called the *Peierls* (or *Peierls–Nabarro*) *stress*. It is not to be confused with the yield stress, for it is known from experiment that individual dislocations can move and give rise to microplasticity under stresses much smaller than the yield stress. Yielding occurs when the applied stress reaches a value such that dislocation sources start to operate and the plastic strain rate, which is proportional to the density and velocity of mobile dislocations (equation 3.13), matches the strain rate imposed by the testing machine (see section 10.5). The Peierls stress arises as a direct consequence of the periodic structure of the crystal lattice and depends sensitively on the form of the force–distance relation between individual atoms, i.e. on the nature of the interatomic bonding. It is a function of the core structure and, for this reason, a unique analytical expression for the Peierls stress cannot be derived. Some important features do emerge, however, from a qualitative assessment.

Dislocation Core Structure

Consider for simplicity a simple cubic crystal. When an extra half-plane of atoms is created by the presence of an edge dislocation of Burgers vector **b**, the atoms in planes above (A) and below (B) the slip plane are displaced by u, as illustrated in Fig. 10.4(a). To accommodate the dislocation, there is a *disregistry* of atomic coordination across the slip plane. It is defined as the *displacement difference* Δu between two atoms on adjacent sites above and below the slip plane i.e. $\Delta u = u(B) - u(A)$. The form of Δu, in units of b,

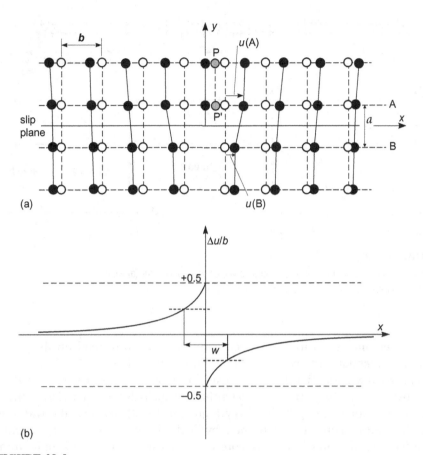

FIGURE 10.4

(a) Displacement of atoms at an edge dislocation. Open and full circles represent the atom positions before and after the extra half-plane is created. (b) Displacement difference Δu across the slip plane for the dislocation in (a). The width of the core in the slip plane is w.

versus x is shown in Fig. 10.4(b). The *width of the dislocation w* is defined as the distance over which the magnitude of the disregistry is greater than one-half of its maximum value, i.e. over which $-b/4 \leq \Delta u \leq b/4$. This parameter is shown in Fig. 10.4(b). The width provides a measure of the size of the dislocation core, i.e. the region within which the displacements and strains are unlikely to be close to the values of elasticity theory.

Typical forms of Δu curves for an edge dislocation in a simple cubic crystal are shown schematically in Fig. 10.5: in these cases, Δu has been increased by b when negative in order to produce continuous curves. The core widths of undissociated dislocations (Figs 10.5(a) and (b)) are usually found by

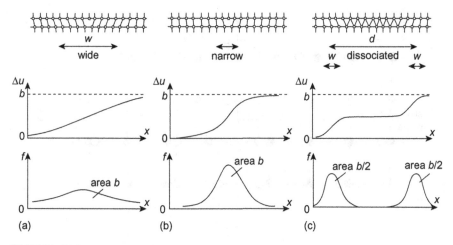

FIGURE 10.5

Atomic positions, disregistry Δu, and Burgers vector distribution f for (a) wide, (b) narrow and (c) dissociated edge dislocations in a simple cubic crystal.

computer simulation to be between b and $5b$, and to depend on the interatomic potential and crystal structure. Dissociation (Fig. 10.5(c)) can only occur when a stable stacking fault exists on the glide plane. The spacing d of the partials may be large and is determined principally by their elastic interaction and the stacking fault energy (equation (5.6)), whereas the width w of the partial cores is determined by inelastic atom–atom interactions. Another useful representation of core structure can be obtained simply from the derivative of the displacement difference curve:

$$f(x) = d(\Delta u)/dx \qquad (10.11)$$

The form of $f(x)$ for the three core structures of Fig. 10.5 is shown by the lower curve in each case. This function is known as the *distribution of Burgers vector* because the area under an $f(x)$ curve equals b. As can be seen from Fig. 10.5, the distribution curve shows clearly where the disregistry is concentrated: it is particularly useful when a dislocation consists of two or more closely-spaced partials. For non-planar cores in which the disregistry and Burgers vector are not distributed mainly on one plane – a situation sometimes found for screw dislocations (sections 6.2 and 6.3) – a two-dimensional plot in which the displacement difference is shown by the length of arrows can be illuminating. This procedure is illustrated for the core of the ⟨111⟩ screw dislocation in a body-centered cubic metal in Figs 6.9 and 6.10. The displacement difference and Burgers vector distribution functions have proved to be particularly useful for presentation of details of the atomic structure of dislocation cores obtained by computer simulation (section 2.4).

The Peierls Barrier

The core disregistry imparts upon a dislocation a core energy and resistance to movement which are functions of the forces between atoms in the core region. In the first estimates of the lattice resistance by Peierls in 1940 and Nabarro in 1947, it was assumed that the atom planes A and B (Fig 10.4) interact by a simple sinusoidal force relation analogous to equation (1.4), and that in equilibrium the resulting disregistry forces on A and B are balanced by the elastic stresses from the two half-crystals above and below these planes. This condition provided an analytical solution for Δu from which w was found to be $a/(1-\nu)$ for an edge dislocation and a for a screw dislocation, where a is the interplanar spacing and ν is Poisson's ratio: the core is therefore 'narrow'. The dislocation energy was also found by combining the disregistry energy, calculated from Δu and the sinusoidal forces, with the elastic energy stored in the two half-crystals. It is similar in form to equation (4.22), with r_0 replaced by approximately $w/3$. When the dislocation in Fig. 10.4 moves to the right to PP', say, the atoms in planes A and B cease to satisfy the equilibrium distribution of Δu, and the disregistry energy increases. Peierls and Nabarro calculated the dislocation energy per unit length as a function of dislocation position, and found that it oscillates with period $b/2$ and maximum fluctuation (known as the *Peierls energy*) given by

$$E_p = \frac{Gb^2}{\pi(1-\nu)}\exp\left(\frac{-2\pi w}{b}\right)$$
(10.12)

The maximum slope of the periodic energy function is the critical force per unit length required to move the dislocation through the crystal. Dividing this by b therefore gives the Peierls stress

$$\tau_p = \frac{2\pi}{b^2}E_p = \frac{2G}{(1-\nu)}\exp\left(\frac{-2\pi w}{b}\right)$$
(10.13)

This simple model successfully predicts that τ_p is orders of magnitude smaller than the theoretical shear strength (equation (1.5)).

Although the Peierls model has now been superseded by more realistic approaches using computer simulation, it has qualitative features generally recognized to be valid. Slip usually occurs on the most widely-spaced planes, and a wide, planar core tends to produce low values of τ_p. For this reason, edge dislocations are generally more mobile than screws (section 3.5). Also, the Peierls energy can be very anisotropic, and dislocations will then tend to lie along the most closely-packed directions, for which τ_p is a maximum, as in Fig. 8.8. Experimental studies of microplasticity and internal friction show that the magnitude of the stress at which dislocations start to move depends sensitively on the nature of the interatomic forces. It is low ($\lesssim 10^{-6}$ to 10^{-5} G) for the face-centered cubic and basal-slip hexagonal metals, in which dislocations dissociate, but is high ($\sim 10^{-2}$ G) for covalent crystals such as silicon and diamond, in which dislocations have a preference for the $\langle 110 \rangle$ orientations either parallel or at $60°$ to their $\frac{1}{2}\langle 110 \rangle$ Burgers vector. The body-centered cubic metals, for which the $\langle 111 \rangle$ screw does not have the planar form of the Peierls model (see section 6.3), and the prism-slip hexagonal metals (section 6.2) fall between these extremes. There is a discrepancy between the experimental values and τ_p which, as defined above, refers to a straight dislocation in a crystal at 0 K. Atomic-scale computer simulation of metals gives τ_p values orders of magnitude higher than those obtained by experiment. Real crystals cannot be cooled to 0 K and dislocations in them are not straight at the atomic scale. These factors assist dislocations to glide at much lower stress than τ_p, a feature confirmed by simulation. They are considered next.

FIGURE 10.6

Schematic illustration of the potential energy surface of a dislocation line due to the Peierls barrier E_p. (After Seeger, Donth and Pfaff, Disc. Faraday Soc. 23, 19, 1957.)

Kink Mechanism

The energy of a dislocation as a function of position in the slip plane is illustrated in Fig. 10.6 for a dislocation lying predominantly parallel to the z-direction, which is a direction corresponding to a low core energy, e.g. one densely packed with atoms. The energy E_0 per unit length is given approximately by equations (4.22) and (4.23), but, as discussed above, superimposed on this is the fluctuation E_p (usually $\ll E_0$) due to the Peierls energy: it has a period a given by the repeat distance of the lattice in the x-direction. If the dislocation is unable to lie entirely in one energy minimum, it contains kinks where it moves from one minimum to the next, as shown in Fig. 10.6. The shape and length m of a kink depend on the value of E_p, and result from the balance of two factors (Fig. 10.7): (i) the dislocation tends to lie as much as possible in the position of minimum energy (this factor alone gives $m = 0$ as for line A) and (ii) the dislocation tends to reduce its energy by being as short as possible (this favors the straight line B with $m \gg a$). In practice, the shape falls between these extremes

FIGURE 10.7

Shape of dislocations running almost parallel to a Peierls energy minimum. The energy per unit length of the dislocation is a minimum along the dashed lines and varies periodically at right angles to the lines. The shape of the dislocation (curve C) is somewhere between the extremes A and B.

(line C), with high E_p producing low m and vice versa. The stress required to move a kink laterally along the line, and thus move the line from one energy minimum to the next, is less than τ_p, which is the stress required to move a long straight line rigidly over the energy hump E_p. Thus, at a low applied stress, pre-existing kinks can move laterally until reaching the nodes at the ends of the dislocation segments. The resulting plastic strain ('preyield microplasticity') is small and leaves long segments of line lying along energy minima. At 0 K, a stress of at least τ_p is therefore required for further dislocation glide. As the temperature is raised, however, there is an increasing probability that atomic vibrations resulting from thermal energy may enable the core to bulge from one minimum to the next over only part of the line (Fig. 10.8) and thus reduce the flow stress. This process of *double-kink* (or *kink-pair*) *nucleation* increases the line length and energy, however. Also, the two kinks X and Y are of opposite sign, i.e. they have the same **b** but approximately opposite line direction, and therefore tend to attract and annihilate each other. As a result, the pair of kinks is not stable under an applied stress unless the length of the bulge, i.e. the spacing XY, is sufficiently large, typically $\simeq 20b$ and $\gg m$. Kink-pair nucleation therefore has an activation energy which is a function of E_p.

FIGURE 10.8

The process of kink-pair nucleation.

The effect on the flow stress is as described in section 10.2. The flow stress for a given applied strain rate decreases with increasing temperature up to T_c as thermal activation becomes increasingly significant. Stress and strain rate at a given temperature are related by equation (10.4) with the Gibbs free energy of activation ΔG^* given by the empirical equation (10.10). The constants p and q in ΔG^* are found to be approximately 3/4 and 4/3 respectively, and ΔF is the energy to form two well-separated kinks. It contains

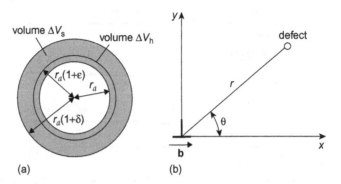

FIGURE 10.9

(a) Elastic model for a defect of natural radius $r_a(1 + \delta)$ inserted in a hole of radius r_a. The final radius is $r_a(1 + \varepsilon)$. (b) Geometry for the interaction of a defect with a dislocation lying along the z-axis.

both elastic and core contributions, and it increases with E_p, but not in a simple way. It is of the order of 0.1 Gb^3 ($\simeq 1 - 2$ eV) for silicon, germanium and for $\langle 111 \rangle$ screw dislocations in the body-centered cubic transition metals, but is less than 0.05 Gb^3 (≤ 0.2 eV) for the face-centered cubic metals. The effect of lattice resistance on the flow stress in the latter case is therefore only significant at very low temperatures.

10.4 INTERACTION BETWEEN POINT DEFECTS AND DISLOCATIONS

Point defects, i.e. vacancies, self-interstitials, and substitutional and interstitial impurities, interact with dislocations. The most important contribution to the interaction between a point defect and a dislocation is usually that due to the distortion the point defect produces in the surrounding crystal. The distortion may interact with the stress field of the dislocation to raise or lower the elastic strain energy of the crystal. This change is the *interaction energy* E_I. If the defect occupies a site where E_I is large and negative, work $|E_I|$ will be required to separate the dislocation from it. This situation may be met either when a dislocation glides through a crystal containing defects or by annealing at sufficiently high temperatures to enable defects to diffuse to such favored positions. Whatever the cause, an increased stress will be required for slip and the crystal will be stronger, as discussed in sections 10.5 and 10.6. In this section, the form of E_I known as the Cottrell–Bilby formula is derived.

The simplest model of a point defect is an elastic sphere of natural radius $r_a(1 + \delta)$ and volume V_s, which is inserted into a spherical hole of radius r_a and volume V_h in an elastic matrix (Fig. 10.9(a)). The sphere and matrix are isotropic with the same shear modulus G and Poisson's ratio ν. The difference between the defect and hole volumes is the *misfit volume* V_{mis}:

$$V_{mis} = V_s - V_h = \frac{4}{3}\pi r_a^3(1 + \delta)^3 - \frac{4}{3}\pi r_a^3 \simeq 4\pi r_a^3 \delta \ (if \ \delta \ll 1) \quad (10.14)$$

The *misfit parameter* δ is positive for oversized defects and negative for undersized ones. On inserting the sphere in the hole, V_h changes by ΔV_h to leave a final defect radius $r_a(1 + \varepsilon)$. The change ΔV_h is given by

$$\Delta V_h = \frac{4}{3}\pi r_a^3(1 + \varepsilon)^3 - \frac{4}{3}\pi r_a^3 \simeq 4\pi r_a^3 \varepsilon \quad (if \ \varepsilon \ll 1) \quad (10.15)$$

Parameter ε is determined by the condition that in the final state the inward and outward pressures developed on the sphere and hole surfaces are equal. The condition is

$$\Delta V_h = \frac{(1 + \nu)}{3(1 - \nu)} V_{mis}$$

i.e.

$$\varepsilon = \frac{(1 + \nu)}{3(1 - \nu)} \delta \quad (10.16)$$

The total volume change experienced by an infinite matrix is ΔV_h, for the strain in the matrix is pure shear with no dilatational part (section 8.4). In a finite body, however, the requirement that the outer surface be stress-free results in a total volume change given by

$$\Delta V = \frac{3(1 - \nu)}{(1 + \nu)} \Delta V_h \quad (10.17)$$

which, from equation (10.16), equals V_{mis}.

If the material is subjected to a pressure $p = -\frac{1}{3}(\sigma_{xx} + \sigma_{yy} + \sigma_{zz})$, the strain energy is changed by the presence of the point defect by

$$E_I = p\Delta V \quad (10.18)$$

For a dislocation, p is evaluated at the site of the defect. For a *screw* dislocation, $p = 0$ (section 4.3) and thus $E_I = 0$. However, for an *edge* dislocation lying along the z-axis (Fig. 10.9(b)), equations (4.17), (10.14) and (10.17) give

$$E_I = \frac{4(1 + \nu)Gbr_a^3\delta}{3(1 - \nu)} \frac{y}{(x^2 + y^2)} \quad (10.19)$$

or, in cylindrical coordinates with $r = (x^2 + y^2)^{1/2}$ and $\theta = \tan^{-1}(y/x)$,

$$E_I = \frac{4(1 + \nu)Gbr_a^3 \delta}{3(1 - \nu)} \frac{\sin \theta}{r} \qquad (10.20)$$

In terms of ε (equation (10.16)), E_I simplifies to

$$E_I = 4Gbr_a^3 \varepsilon \frac{\sin \theta}{r} \qquad (10.21)$$

but note that this equation would not be valid if the defect were, say, incompressible, for then equation (10.16) would be replaced by $\varepsilon = \delta$.

For an oversized defect ($\delta > 0$) E_I is positive for sites above the slip plane ($0 < \theta < \pi$) and negative below ($\pi < \theta < 2\pi$). This is because the edge dislocation produces compression in the region of the extra half-plane and tension below (section 4.3). The positions of attraction and repulsion are reversed for an undersized defect ($\delta < 0$). The value of r_a in equation (10.20) can be estimated from the fact that V_h is approximately equal to the volume per atom Ω for substitutional defects and vacancies, and is somewhat smaller for interstitial defects. A given species of defect produces strain in the lattice proportional to δ, and this parameter can be determined by measurement of the lattice parameter as a function of defect concentration. It is found to range from about -0.1 to 0 for vacancies, -0.15 to $+0.15$ for substitutional solutes, and 0.1 to 1.0 for interstitial atoms. The sites of minimum E_I, i.e. maximum *binding energy* of the defect to the dislocation, are at $\theta = \pi/2$ or $3\pi/2$, depending on the sign of δ. The strongest binding occurs within the dislocation core at $r \approx b$, and from equation (10.20) the energy is approximately $G\Omega|\delta|$. It ranges from about $3|\delta|$ eV for the close-packed metals to $20|\delta|$ eV for silicon and germanium. Although the use of equation (4.17) for p within the dislocation core is questionable, these energy values may be considered upper limits and are within an order of magnitude of estimates obtained from experiment and computer simulation.

The interaction of a dislocation with a spherically symmetric defect is a special case of a more-general *size-effect* interaction. Many defects occupy sites having lower symmetry than the host crystal and thereby produce asymmetric distortions. Unlike the spherical defect, the asymmetric defect interacts with both hydrostatic and shear stress fields, and therefore interacts with edge *and* screw dislocations. A well-known example of such a defect is the interstitial solute atom in a body-centered cubic metal, e.g. carbon in α-iron. The defect geometry is illustrated in Fig. 10.10. The most favorable site is the *octahedral* site at the center of a cube face or, equivalently, at the center of a cube edge. The interstitial has two nearest-neighbor atoms E and F a distance $a/2$ away and four second-nearest neighbors A, B, C and D a distance $a/\sqrt{2}$

away. An interstitial atom with a diameter larger than $(a - 2R)$, where R is the radius of an iron atom, produces a tetragonal distortion by displacing atoms E and F more strongly than the second-nearest neighbors. The misfits in the x, y, z-directions for carbon in iron are estimated to be

$$\delta_{xx} = \delta_{yy} = -0.05; \quad \delta_{zz} = +0.43 \qquad (10.22)$$

The interaction energy with a dislocation is, by analogy with equation (10.18)

$$E_I = -(\sigma_{xx}\delta_{xx} + \sigma_{yy}\delta_{yy} + \sigma_{zz}\delta_{zz})\frac{4}{3}\pi r_a^3 \qquad (10.23)$$

FIGURE 10.10
Octahedral interstitial site in a body-centered cubic cell.

where σ_{xx}, σ_{yy} and σ_{zz} are the stresses produced by the dislocation. For the screw dislocation, which lies along $\langle 111 \rangle$, the magnitude of E_I is comparable with that for the edge. In both cases E_I is proportional to r^{-1} as in equation (10.20), but has a more complicated orientation dependence than $\sin \theta$.

A second form of elastic interaction arises if the defect is considered to have different elastic constants from the surrounding matrix. Point defects can increase or decrease the elastic modulus. A vacancy, for example, is a soft region of zero modulus. Both hard and soft defects induce a change in the stress field of a dislocation, and this produces the *inhomogeneity* interaction energy. The interaction is always attractive for soft defects, because they are attracted to regions of high elastic energy density which they tend to reduce. (This is analogous to the attraction of dislocations to a free surface (section 4.8)). Hard defects are repelled. The dislocation strain energy over the defect volume is proportional to $1/r^2$ for both edges and screws (see equations (4.9), (4.15) and (4.16)), and the interaction energy therefore decreases as $1/r^2$. Although it is of second order in comparison with the $1/r$ size-effect interaction, it can be important for substitutional atoms and vacancies when the misfit parameter δ is small.

There are sources of interaction between point defects and dislocations which are not included in the elastic model. One is the *electrical interaction*. In *metals*, the conduction electron density tends to increase in the dilated region below the half-plane of the edge dislocation and decrease in the compressed region. The resulting electric dipole could, in principle, interact with a solute atom of different valency from the solvent atoms, but free-electron screening is believed to reduce the interaction to negligible proportions in comparison with the size and inhomogeneity effects. In *ionic* crystals, lack of electron screening leads to a more significant interaction, but of more importance is the presence of charged jogs. The electric charge associated with jogs (section 6.4) produces a $1/r$ Coulomb electrostatic interaction with vacancies and charged impurity ions, and this probably dominates the total interaction energy in such materials. In *covalent* crystals, the relative importance of the variety of possible effects is less clearly understood. In addition to the elastic interaction and the electric-dipole interaction with impurities of different valency, an electrostatic-interaction can occur if conduction electrons are captured by the dangling bonds which may exist in some dislocation cores (section 6.6).

Finally, when a perfect dislocation dissociates into partial dislocations, the ribbon of stacking fault formed changes the crystal structure locally. For example, the faulted stacking sequence is hexagonal-close-packed in face-centered cubic metals (section 5.3) and face-centered cubic in the hexagonal metals (section 6.2). Thus, the solute solubility in the fault region may be different from that in the surrounding matrix. The resulting change in chemical potential will cause solute atoms to diffuse to the fault. This *chemical* (or *Suzuki*) *effect* is not a long-range interaction, but it may present a barrier to the motion of dissociated dislocations.

10.5 SOLUTE ATMOSPHERES AND YIELD PHENOMENA

Dislocation Locking

In a crystal containing point defects in solution, the energy of a defect in the vicinity of a dislocation is changed by E_I, given by the sum of the terms discussed above. The equilibrium defect concentration c at a position (x, y) near the dislocation therefore changes from c_0, the value a long way from the dislocation, to

$$c(x, y) = c_0 \exp[-E_I(x, y)/kT] \tag{10.24}$$

(This formula assumes the concentration c is weak, i.e. fewer than half the core sites are occupied, and that the defects do not interact with each other.) Defects therefore tend to congregate in core regions where E_I is large and negative, and dense atmospheres of solute atoms can form in otherwise weak solutions with small values of c_0. The condition for formation of these *Cottrell atmospheres* is that the temperature is sufficiently high for defect migration to occur but not so high that the entropy contribution to the free energy causes the atmosphere to 'evaporate' into the solvent matrix. It is readily shown from equation (10.24) that even in dilute solutions with, say, $c_0 = 0.001$, dense atmospheres with $c \gtrsim 0.5$ may be expected at $T = 0.5\ T_m$, where T_m is the melting point, in regions where $-E_I \gtrsim 3kT_m$. This corresponds to a defect-dislocation binding energy of typically 0.2 to 0.5 eV for metals, which is fairly commonplace for interstitial solutes.

The kinetics of the migration of defects to dislocations are governed by the drift of defects under the influence of the interaction energy E_I superimposed on the random jumps associated with diffusion. The path taken by a defect depends on the precise form of the angular part of E_I. For example, a misfitting spherical defect interacts with a straight edge dislocation according to the form (equations (10.19), (10.20))

$$E_I = A \frac{y}{(x^2 + y^2)} = A \frac{\sin \theta}{r} \qquad (10.25)$$

so that lines of constant E_I are circles centered on the y-axis as shown in Fig. 10.11. A defect will tend to migrate into the core region by following a path which is everywhere perpendicular to a constant energy contour. Such a flow line is itself a circle which is centered on the x-axis, as indicated by the broken lines in Fig. 10.11.

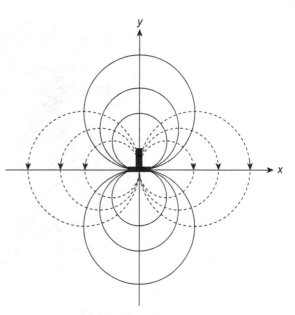

FIGURE 10.11

Equipotential contours (full lines) for the elastic interaction potential given by equation (10.25), between solute atoms and an edge dislocation along the z-axis with Burgers vector parallel to the x-axis. Broken lines are lines of flow for oversized solute atoms migrating to the dislocation.

However, the rate of arrival of defects at the core depends on the exponent of r in E_I, rather than the shape of the flow lines. Solution of the appropriate diffusion equations shows that the rate varies with time t as $t^{2/3}$ for first-order interactions $(E_I \propto r^{-1})$ and $t^{1/2}$ for second-order ones $(E_I \propto r^{-2})$. The former dependence is that obtained by Cottrell and Bilby for carbon solutes in α-iron.

Within the core itself, the arrival of defects can have several effects. Vacancies and self-interstitial atoms can cause climb, as discussed in Chapter 3. Small interstitial dislocation loops formed by radiation damage in metals can migrate to the tensile strain region below the extra half-plane of an edge dislocation. Dislocations decorated in this way in molybdenum irradiated with neutrons are shown in Fig. 10.12(a). In alloys, solute atoms may segregate as shown by phosphorus in the 3DAP images of iron in Fig. 2.17, and solute atoms in high local concentrations may interact to form precipitates along the core, such as those forming carbides in Fig. 10.12(b) and those discussed in the decoration technique in section 2.3. On the other hand, solutes may remain in solution as a dense atmosphere. Both effects can annul the long-range stress field of the dislocation and solute flow then ceases.

The importance of solute segregation in or near the core as far as mechanical properties are concerned, is that an extra stress is required to overcome the attractive dislocation−defect interactions and move the dislocation away from the concentrated solute region. The dislocation is said to be *locked* by the solutes. Once a sufficiently high stress has been applied to separate a

(a)   (b)

FIGURE 10.12

Transmission electron micrographs of dislocations decorated with (a) small interstitial loops in a molybdenum irradiated at 320 K by fast neutrons and (b) carbide precipitate platelets (viewed edge-on) in iron. ((a) From Trinkaus, Singh and Foreman, J. Nucl. Mater. *249, 91, 1997. Copyright (1997) with permission from Elsevier. (b) From Hull and Mogford, Phil. Mag. 6, 535, 1961.*)

dislocation from the defects, its movement is unaffected by locking. However, subsequent heat treatment which allows the defects to diffuse back to the dislocations re-establishes locking: this is the basis of *strain aging*. Also, at sufficiently high deformation temperatures and low strain rates, point defect mobility may enable solutes to repeatedly lock dislocations during dislocation motion. This produces a repeated yielding process known as *dynamic strain aging* (or *Portevin–Le Chatelier effect*), characterized by a serrated stress–strain curve.

The unlocking stress is treated here for the case of a row of misfitting spherical defects lying in the position of maximum binding along the core of a straight edge dislocation, as illustrated in Fig. 10.13. They represent, therefore, a small atmosphere of, say, carbon atoms (or possibly very small coherent precipitates). It is assumed that the core sites are saturated, i.e. that every defect site (of spacing b) along the core is occupied. (Note that even at saturation, the number of solutes per unit volume in such atmospheres $\approx \rho/b$, where ρ is the dislocation density, is only a small fraction of the total number in the crystal $\simeq c_0/b^3$. For example, taking $\rho = 10^6$ mm^{-2} and $b = 0.3$ nm, the ratio $\rho b^2/c_0$ is only $10^{-7}/c_0$.) If the position of the solutes is r_0 $(\simeq b)$

FIGURE 10.13
Diagrammatic representation of a row of solute atoms lying in the position of maximum binding at an edge dislocation. An applied shear stress will cause the dislocation to separate from the solute atoms by gliding in the slip plane to position x.

below the slip plane, as shown, the interaction energy of the dislocation with one defect when it has been displaced to position x in the slip plane is, from equation (10.25)

$$E_I(x) = \frac{A(-r_0)}{(x^2 + r_0^2)} \tag{10.26}$$

The force in the x-direction on the line in this position is

$$K(x) = \frac{-dE_I(x)}{dx} = \frac{-2Ar_0x}{(x^2 + r_0^2)^2} \tag{10.27}$$

The force due to the $1/b$ defects per unit length is $K(x)/b$. Thus the resolved shear stress necessary to displace the line to position x is

$$\tau = -\frac{1}{b^2}K(x) = \frac{2Ar_0x}{b^2(x^2 + r_0^2)^2} \tag{10.28}$$

It has a maximum τ_0 at $x = r_0/\sqrt{3}$ given by

$$\tau_0 = \frac{3\sqrt{3}A}{8b^2r_0^2} \tag{10.29}$$

The form of τ is shown in Fig. 10.14. From equation (10.20), $A \simeq Gb\Omega|\delta|$, so that taking $r_0 = b$, $\tau_0 \simeq 0.2G|\delta|$. This is undoubtedly an upper limit, since it is based on the application of elasticity within the dislocation core.

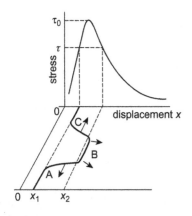

FIGURE 10.14

Separation of a dislocation from a row of condensed solute atoms. Initially the dislocation lies along the line $x = 0$. Under the applied stress τ it moves forward to the position of stable equilibrium x_1. To break away it must reach the position of unstable equilibrium x_2. ABC represents a loop of dislocation, formed by a thermal stress fluctuation, in the process of breaking away. *(From Cottrell (1957), Properties of Materials at High Rates of Strain, Instn. Mech. Eng., London.)*

Nevertheless, it demonstrates that a considerable increase in the yield strength may be expected from concentrated solute atmospheres.

The estimate for τ_0 is the yield stress at 0 K. At non-zero temperatures, thermal activation (section 10.2) may assist unlocking and reduce the applied stress required. The mechanism is illustrated in Fig. 10.14, where a straight dislocation AC under an applied stress τ ($<\tau_0$) is held in stable equilibrium at position x_1 until thermal fluctuations assist part of the line B to overcome the resisting force. The process is analogous to double-kink-pair nucleation over the Peierls barrier (section 10.3), except that $(x_2 - x_1)$ depends on the atmosphere dimensions and is not simply the lattice spacing. The analysis of Cottrell and Bilby gives an activation energy in the form of equation (10.10) with $p = 1$ and $q = 1.5$. The applied stress for unlocking therefore decreases rapidly from τ_0 with increasing temperature. However, as can be deduced from the preceding analysis, the predicted values of barrier energy ΔF for dense atmospheres and core precipitates are frequently too large for thermally-activated unpinning to be responsible for the temperature dependence of the yield stress. In materials in which strong atmospheres form it is probable that only a small fraction of locked dislocations become free, as a result, perhaps, of local stress concentrations. The temperature-dependence of the yield stress is then dominated by other effects, such as the Peierls stress and solution strengthening. Nevertheless, locking may still have a significant influence on the actual yield behavior, as discussed below.

Yield Drop

The presence of strong locking effects can lead in some alloys to a pronounced drop in the stress required for plastic deformation immediately after yielding. The stress–strain curve that results is shown in Fig. 10.1(c). If the deformed specimen is unloaded and reloaded immediately, a yield drop does not accompany the onset of plastic flow a second time. However, if the sample is given a suitable anneal prior to reloading, strain aging causes the yield drop to return. The principles involved in the yield drop phenomenon are as follows.

Consider a tensile test, shown schematically in Fig. 10.15. The elasticity of the machine (including the load cell) and specimen is represented by an imaginary spring. The crosshead moves at a constant speed $s = dl/dt$ so that the crosshead displacement at time t is st. The total elastic displacement of the spring is KF where F is the applied force and K the spring constant. The plastic elongation of the specimen is $\varepsilon_p l_0$ where ε_p is the plastic strain and l_0 is the original gauge length. Thus

$$st = KF + \varepsilon_p l_0 \qquad (10.30)$$

The plastic strain of the specimen is therefore

$$\varepsilon_p = \frac{st - KF}{l_0} \qquad (10.31)$$

and the plastic strain rate is

$$\dot{\varepsilon}_p = \frac{d\varepsilon_p}{dt} = \frac{s - K\frac{dF}{dt}}{l_0} \qquad (10.32)$$

Taking the resolved shear stress τ as $F/2A_0$, where A_0 is the original cross-sectional area of the specimen, equation (10.32) also gives

$$\frac{d\tau}{dt} = \frac{s - \dot{\varepsilon}_p l_0}{2A_0 K} \qquad (10.33)$$

Since $s = dl/dt$, the measured hardening rate

$$\frac{d\tau}{dl} = \frac{1}{2A_0 K}\left(1 - \frac{\dot{\varepsilon}_p l_0}{s}\right) \qquad (10.34)$$

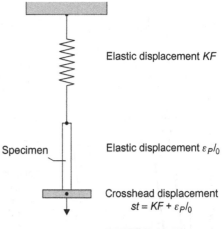

Elastic displacement KF

Specimen

Elastic displacement $\varepsilon_p l_0$

Crosshead displacement
$st = KF + \varepsilon_p l_0$

FIGURE 10.15
Schematic diagram of a tensile machine. The spring represents the elastic properties of the machine and specimen. *(After Johnston and Gilman, J. Appl. Phys. 30, 129, 1959.)*

can be seen to be dependent on the elastic properties of the machine and specimen and the instantaneous plastic strain rate of the specimen.

The shape of the stress–strain curve given by equation (10.34) can be related to the dislocation behavior through the plastic strain rate term $\dot{\varepsilon}_p$ using equation (3.13). However, since both ρ_m and $\bar{v}$ will vary with stress and strain, it is necessary to know the variation of both before any predictions can be made. This information is only available in isolated cases. Johnston (1962) reviewed the effects of all the machine and specimen dislocation variables on the stress–strain curve and some of his results are illustrated in Fig. 10.16. The calculations were based on data obtained from LiF crystals in which the measured dislocation density increases approximately linearly with strain in the early stages and the dislocation velocity can be related to the applied shear stress through a relation of the type (3.3), namely

$$\bar{v} \propto \tau^m \qquad (10.35)$$

where m was taken as 16.5 for LiF.

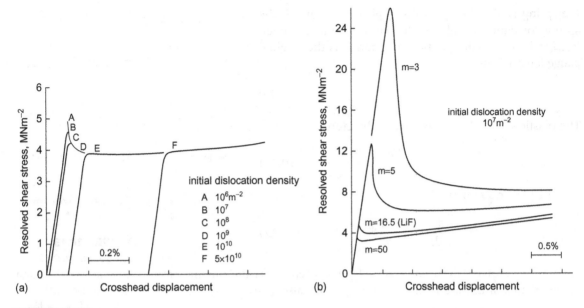

FIGURE 10.16

(a) Effect of initial density of mobile dislocations on the yield point. (b) Effect on the yield point of changing the stress dependence of dislocation velocity for an initial mobile dislocation density of 10^3 cm^{-2}. *(After Johnston, J. Appl. Phys. 33, 2716, 1962.)*

Figure 10.16(a) shows the effect of changing the *initial* density of mobile dislocations from 10^6 to 5×10^{10} m^{-2}($=5 \times 10^6$ cm^{-2}). The line *OA* represents the situation where there is no dislocation movement and $\dot{\varepsilon} = 0$ in equation (10.34), i.e. completely elastic behavior, $d\tau/dl = (2A_0K)^{-1}$. The curves show that a sharp yield drop is obtained for low initial values of ρ_m and that the yield drop decreases as ρ_m increases. The curves D, E, F in Fig. 10.16(a) have been displaced from the origin for convenience. Figure 10.16(b) shows the calculated stress–strain curves for different values of the parameter *m* in relation (10.35). The yielding shows a strong dependence on *m*. These effects are best understood by noting from equation (3.13) that at a constant plastic strain rate

$$\rho_{mu}\bar{v}_u = \rho_{ml}\bar{v}_l \tag{10.36}$$

where subscripts *u* and *l* refer to values at upper and lower yield points respectively (see Fig. 10.1(c)). From relations (10.35) and (10.36), the ratio of the upper to lower yield stresses is

$$\frac{\tau_u}{\tau_l} = \left[\frac{\rho_{ml}}{\rho_{mu}}\right]^{1/m} \tag{10.37}$$

The ratio is therefore largest for small m and/or large (ρ_{ml}/ρ_{mu}). The value of m varies considerably from one material to another (section 3.5). In α-iron, for example, where it is large ($m \simeq 40$), equation (10.37) shows that ρ_m has to increase by 10^4 times at yield for the stress to drop by 20 per cent. That is, if $\rho_{ml} = 10^7 \text{ cm}^{-2}$, $\rho_{mu} = 10^3 \text{ cm}^{-2}$. Since the *total* dislocation density initially might be typically 10^6 to 10^7 cm^{-2}, locking plays an important role in ensuring that the initial mobile density ρ_{mu} is small.

The physical interpretation of this analysis is as follows. When a crystal containing a low density of dislocations which are free to glide is strained using a constant crosshead speed s, the dislocations cannot move fast enough at low stresses to produce sufficient strain in the specimen. The stress in the specimen (and spring K) therefore rises, and as it does so the dislocations multiply rapidly and move faster, as implied by relation (10.35). The stress stops rising (point E on Fig. 10.1(c)) when $d\tau/dl = 0$, which from equation (10.34) occurs when $\dot{\varepsilon}_p = b\rho_m\bar{v} = s/l_0$, i.e. the strain rate of the specimen equals the applied strain rate. However, multiplication continues with increasing strain, producing more than enough dislocations and $b\rho_m\bar{v} > s/l_0$. The stress therefore drops until $\bar{v}$ is slow enough for the two strain rates to match. In general, three factors favor this phenomenon. First, the initial density of mobile dislocations ρ_{mu} must be small; second, the dislocation velocity must not increase too rapidly with increasing stress, i.e. small m; and third, the dislocations must multiply rapidly. The first is the most important.

Irradiation Hardening and Dislocation Channeling

A phenomenon that illustrates several features discussed in this section is found to occur in metals irradiated with energetic atomic particles, e.g. fast neutrons in a fission reactor. The incident particles cause *radiation damage* by displacing many atoms from their lattice sites by collisions, thereby creating supersaturations of vacancies and self-interstitial atoms. Many collect together, forming interstitial and vacancy dislocation loops or vacancy stacking-fault tetrahedra (sections 3.7, 5.5, 5.7, 6.2 and 6.3), and others diffuse to dislocations, causing climb, or to other defect sinks such as grain boundaries. The loops and small clusters that remain in the lattice act as obstacles to dislocation motion in a similar way to that described for alloys in the next section, and raise the stress required for yield and subsequent plastic flow, as seen in Fig. 10.17(a) and (b) for proton-irradiated single crystals of copper and neutron-irradiated polycrystalline iron, respectively. This is *irradiation hardening*.

As noted in section 6.3, small loops consisting of clusters of self-interstitials may be quite mobile if they have perfect Burgers vectors and so they will be

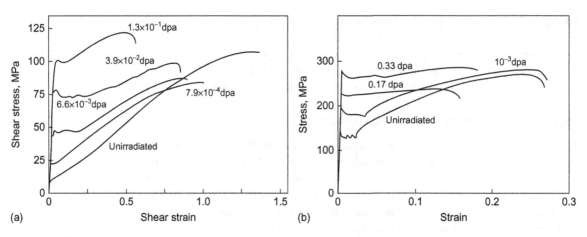

FIGURE 10.17

Stress—strain curves for (a) copper single crystals ($\langle 110 \rangle$ orientation) irradiated by 590 MeV protons and (b) polycrystalline iron irradiated by fast neutrons to different doses at room temperature. The damage dose is given as the average number of times each atom is displaced by the radiation (displacements per atom (dpa)). *(Reprinted from Victoria, Baluc, Bailat, Dai, Luppo, Schaublin and Singh, J. Nucl Mater. 276, 114, 2000, with permission from Elsevier Science.)*

able to migrate to attractive regions where their energy is lowered just below the core of edge dislocations. They can form atmospheres, as shown in molybdenum in Fig. 10.12(a), and so lock the dislocations and cause a yield drop in metals where it is not usually observed, e.g. copper at high irradiation dose as in Fig. 10.17(a). In materials where the density of defects is high, dislocation motion after yielding may create regions almost free of obstacles, as seen in the example of Fig. 10.18. The mechanisms involved in this *dislocation channeling* are not understood, but may be associated with the ability of gliding dislocations to absorb by climb the self-interstitial atoms and vacancies in the clusters they encounter during glide, so that channeling can occur by repeated glide and climb.

10.6 THE FLOW STRESS FOR RANDOM ARRAYS OF OBSTACLES

The ability to confer increased strength on a crystal of one element by the deliberate addition of atoms of other elements is one of the important achievements of materials technology. It is possible to raise the yield strength of metals such as aluminum, copper and nickel, for example, up to 100 or more times the value for the pure elements by suitable choice of alloying additions and heat treatment. Clearly, alloying introduces barriers to the motion of dislocations, but the mechanism by which dislocations are hindered depends on the form the solute atoms adopt within the solvent lattice. These microstructural features are assessed in section 10.7. In the present section, general expressions for the flow stress at 0 K are derived, and it is sufficient to know that the obstacles exert forces on a dislocation as it moves on its slip plane. The theory of obstacle strengthening can be complicated but the important features can be distinguished with the simple approach adopted here.

An obstacle may be classed as *strong* or *weak* depending on whether or not a dislocation bends through a large angle in its vicinity. Also, an obstacle may be *localized* or *diffuse* depending on whether or not the force it exerts on the line is confined to a small part of the line or is distributed approximately uniformly over a long length. The latter situation is considered first.

FIGURE 10.18

Transmission electron micrograph of dislocation channeling in a palladium single crystal irradiated by 590 MeV protons to a damage dose of 0.12 dpa. The sample contains a high density of point defect clusters at this dose and when yielding occurs channels parallel to {111} slip planes are created that are almost free of defects. *(Reprinted from Victoria, Baluc, Bailat, Dai, Luppo, Schaublin and Singh, J. Nucl. Mater. 276, 114, 2000, with permission from Elsevier Science.)*

Diffuse Forces

Each obstacle is assumed to create an internal shear stress τ_i in the matrix proportional to the misfit parameter δ (sections 8.4 and 10.4). This produces a force on a dislocation of maximum value $\tau_i b$ per unit length. It is further assumed that τ_i is constant over a region of size Λ, the spacing of the obstacles. An obstacle can therefore bend a dislocation against the opposition of its line tension to a radius of curvature R given by (equation (4.30))

$$R = \alpha G b/\tau_i \simeq G b/2\tau_i \tag{10.38}$$

Strong forces result in $R \lesssim \Lambda$ and weak ones in $R \gg \Lambda$. *Strong* forces therefore bend the line sufficiently for it to follow the contours of minimum interaction energy with each obstacle, as shown in Fig. 10.19(a), and the flow stress is that required to overcome the mean internal stress:

$$\tau_{\text{diff}}(\text{strong}) \simeq \frac{2}{\pi}\tau_i \tag{10.39}$$

Line tension prevents the dislocation bending round individual *weak* obstacles, and the dislocation samples regions of both high and low interaction energy, trying to avoid the former (Fig. 10.19(b)). During plastic flow, the

Dislocation motion
(a)

Dislocation motion
(b)

FIGURE 10.19

Dislocation on a slip plane in which obstacles of mean spacing Λ exert diffuse forces. It advances from the full to the dashed line shape as indicated. The obstacle forces are strong in (a) and weak in (b). *(After Nabarro (1975).)*

line advances by segments of length L ($\gg \Lambda$) moving independently of each other from one position of low energy to another. Over the length L, there are $n \simeq L/\Lambda$ obstacles, and in a random array, there will be an excess $\sqrt{n}$ acting in one direction on the line. The flow stress is therefore that required for the segment L to overcome the mean internal stress of $\sqrt{n}$ obstacles, each of which acts over length Λ, i.e.

$$\tau b L \simeq \sqrt{n}\frac{2}{\pi}\tau_i b\Lambda$$

therefore

$$\tau \simeq \frac{2\tau_i}{\pi}\left(\frac{\Lambda}{L}\right)^{1/2} \tag{10.40}$$

It has been recognized from the time of the earliest theories that it is difficult to estimate L. Here, it is assumed that the segment L has radius of curvature R ($\gg L$) equal to $Gb/2\tau$, and that its midpoint moves a distance Λ at each advance forward. Then, from the geometry of a circle:

$$\frac{\Lambda}{2} = R - [R^2 - (L/2)^2]^{1/2} \simeq \frac{L^2}{8R} \tag{10.41}$$

Replacing R by $Gb/2\tau$ to find L and substituting in equation (10.40), the flow stress to move a dislocation through an array of weak, diffuse forces is found to be

$$\tau_{\text{diff}}(\text{weak}) \simeq \frac{2}{\pi^{4/3}}\tau_i\left[\frac{\tau_i\Lambda}{Gb}\right]^{1/3} \tag{10.42}$$

Other assumptions change both the numerical constant and the exponent in this expression.

Localized Forces

In this situation, the dislocation experiences no internal stress in regions between the obstacles, and the resistance offered by each obstacle is represented by a force K acting at a point on the line. Under an applied shear stress τ, the line bows between neighboring obstacles to a radius of curvature $R \simeq Gb/2\tau$, as shown in Fig. 10.20(a). From equilibrium between the line tension forces T at an obstacle and K:

$$K = 2T\cos\phi \simeq Gb^2\cos\phi \tag{10.43}$$

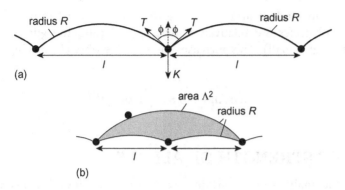

FIGURE 10.20
(a) Dislocation on a slip plane in which each obstacle exerts a localized glide resistance force K, balanced in equilibrium by line tension forces T. (b) The area used to calculate the effective obstacle spacing l for weak forces.

If the obstacle spacing is l (Fig. 10.20(a)), each obstacle resists the forward force τbl on a segment of line of initial length l. The dislocation breaks away from an obstacle when τbl equals the maximum resisting force K_{max}. The flow stress for a regular square array of localized obstacles is therefore

$$\tau_{loc} = \frac{K_{max}}{bl} = \frac{2T}{bl}\cos\phi_c \simeq \frac{Gb}{l}\cos\phi_c \qquad (10.44)$$

where the obstacle strength is characterized by the critical angle $\phi_c = \cos^{-1}(K_{max}/2T)$ at which breakaway occurs.

In random arrays of *weak* obstacles, for which $\phi_c \simeq \pi/2$, a dislocation advances by breaking from a few widely-spaced obstacles before unpinning along the entire length. It retains a roughly straight shape. The effective obstacle spacing l along an almost straight line is given by the *Friedel relation*, as follows. In steady state at the flow stress, when a dislocation unpins from one obstacle it moves forward to encounter one other. The area swept in so doing is therefore the area of glide plane per obstacle Λ^2, where Λ is the average spacing of obstacles in the plane. From Fig. 10.20(b), this area is simply the area of one large segment of a circle of radius R minus that of two others, and is

$$\Lambda^2 \simeq \frac{2l^3}{3R} - \frac{l^3}{6R} \ (if \ l \ll R) \simeq \frac{l^3\tau}{Gb} \qquad (10.45)$$

From equation (10.44) the effective obstacle spacing l is therefore $\Lambda/(\cos\phi_c)^{1/2}$, so that using the same equation, the flow stress for weak localized obstacles is

$$\tau_{loc}(weak) \simeq \frac{Gb}{\Lambda}(\cos\phi_c)^{3/2} \qquad (10.46)$$

Dislocations in random arrays of *strong* obstacles ($\phi_c \simeq 0$) move in regions where the obstacles are relatively widely spaced (and the glide resistance correspondingly small). This increases the effective obstacle spacing and the flow stress is

$$\tau_{\text{loc}}(\text{strong}) \simeq 0.84 \frac{Gb}{\Lambda}(\cos \phi_c)^{3/2} \qquad (10.47)$$

10.7 THE STRENGTH OF ALLOYS

To apply the analyses of the previous section to real alloys, it is necessary to consider the relative values of the flow stress given by equations (10.39), (10.42), (10.46) and (10.47). First, however, a simplified account of the effects of alloying in obstacle structure is presented.

Solutions, Precipitates and Aging

There is usually a maximum concentration to which the atoms of element B can be dissolved in solution in a crystal of element A. For concentrations below this *solubility limit*, any strengthening effect is due to the resistance to dislocation motion created by the atoms of B dispersed randomly in the A lattice. Important examples include alpha-brass (Cu alloyed with Zn), the titanium alloy Ti−6A1−4V and the 5000 series aluminum alloys (Al−Mg).

For concentrations of solute above the solubility limit, the excess B atoms tend to precipitate within particles of either element B or a compound of A and B. The solubility limit varies with temperature, however. It is therefore possible to have an alloy composition which is less than the limit at one temperature but exceeds it at another. An example afforded by the aluminum−copper system is shown by the equilibrium phase diagram in Fig. 10.21. The α-phase is a *solid solution* of copper atoms dissolved substitutionally in aluminum. At 550°C, the solubility limit is almost 6 wt per cent, but at room temperature it is less than 0.1 wt per cent. Thus, the alloy 'Duralumin', which contains 4 wt per cent copper, is a homogeneous solid solution at the *solution treatment* temperature of 550°C. But on slowly cooling the alloy, the second phase starts to precipitate out at the temperature indicated by point x, and at 20°C the alloy consists of large $CuAl_2$ precipitates (θ-phase) in equilibrium with the aluminum-rich matrix. However, if the solution-treated alloy is quenched to room temperature, insufficient time is allowed for precipitation and a *super-saturated* solid solution is obtained. On subsequent annealing at 150°C, say, there is just sufficient

FIGURE 10.21

Section of the aluminum−copper equilibrium diagram.

thermal energy available for precipitate *nucleation* to occur. Initially, a fine dispersion of numerous small precipitates is formed, but with increasing aging time at the elevated temperature, the dispersion coarsens. The number of precipitates decreases and their size and spacing increases, and eventually, after very long times, the equilibrium structure is established.

○ Al   ● Cu

FIGURE 10.22

Schematic representation of a {100} section through a GP zone in aluminum–copper illustrating the coherency strains. The zones in real alloys are larger than that shown here.

In many alloys, such as Al–Cu, metastable phases occur during the transition of the precipitates to equilibrium. The structure of the interface between these zones and the matrix governs the kinetics of their formation and growth. The first zones to form are small and *coherent*, in the sense defined in section 9.6. In Al–Cu, for example, the copper atoms nucleate platelets (known as GP zones) of about 10 nm diameter on the {100} aluminum planes, as shown schematically in Fig. 10.22. Since the copper atoms are smaller than the aluminum atoms, the surrounding matrix is strained, as indicated. (In some other alloys, the zones are oversized.) As the zones grow by bulk diffusion, they thicken (θ''-phase) and the large elastic energy associated with the coherency strains is reduced by the transition to semicoherent zones (θ'-phase) with misfit dislocations (section 9.6) in the interface. These precipitates have crystal structures which may be coherent with the matrix on some faces but not all. Eventually, after further growth, coherency is lost completely and precipitates of the equilibrium phase – $CuAl_2$ (θ-phase) in an Al–Cu alloy – are produced. Aging therefore results in a transition from (a) a solid solution of misfitting atoms to (b) a fine dispersion of coherent precipitates surrounded by elastic strain to (c) a coarse dispersion of particles with incoherent interfaces and negligible lattice strain. These structures offer differing resistance to dislocation motion, and so the yield strength of the alloy changes with time, as illustrated in Fig. 10.23. The solid solution is stronger than the pure metal, but by aging it is possible to further increase the strength. A peak strength is encountered corresponding to a critical dispersion of coherent or semicoherent precipitates, and beyond this the strength falls as *overaging* occurs. Selection of the optimum heat-treatment conditions is therefore crucial to maximizing the strength of alloys. Important examples include the aluminum alloy series 2000 (Al–Cu), 6000 (Al–Mg–Si) and 7000 (Al–Zn).

The behavior of a dislocation under applied stress is different in the three situations (a)–(c) described above, and the flow stress for the three cases is treated separately below using the analysis of section 10.6.

Solution Strengthening

In the *diffuse*-force model of a solid solution, there are c/b^3 obstacles per unit volume, where b is the interatomic spacing and c the atomic-fraction

Critical dispersion
of coherent or semicoherent
precipitates

Homogeneous
solid solution

Incoherent
precipitates

Yield stress

Aging time
(particle size)

FIGURE 10.23

Variation of yield stress
with aging time typical of
an age-hardening
aluminum alloy.

concentration of solute, so that the mean spacing of neighboring solute atoms is

$$\Lambda = b/c^{1/3} \qquad (10.48)$$

The maximum internal stress averaged over the space of radius $\Lambda/2$ around each solute is

$$\tau_i \simeq G|\delta|c \ln(1/c) \qquad (10.49)$$

In the *local*-force model, on the other hand, only those solute atoms in the two planes immediately adjacent to the slip plane contribute. There are $2c/b^2$ of them per unit area of plane, so that in this case

$$\Lambda = b/(2c)^{1/2} \qquad (10.50)$$

The maximum resisting force exerted by each of these solute atoms is found from equation (10.27) (and the subsequent discussion) to be

$$K_{max} \simeq \frac{1}{5}Gb^2|\delta| \qquad (10.51)$$

The quantities τ_i and K_{max} are seldom large enough for strong-force conditions to apply. Equations (10.42) and (10.46) based on weak forces are therefore appropriate, and the latter gives much the largest flow stress when the parameters of equations (10.48) to (10.51) are substituted. The flow stress for a solid solution is thus

$$\tau_{loc}(\text{weak}) \simeq \sqrt{2}Gc^{1/2}(\cos \phi_c)^{3/2} \qquad (10.52)$$

where $\cos \phi_c \simeq K_{max}/Gb^2$ is given by equation (10.51). The stress increases as the square root of the concentration. For the largest misfit expected for substitutional solutes, i.e. $|\delta| \simeq 0.15$ giving $\cos \phi_c \simeq 0.03$, the flow stress will be raised to about $G/400$ when $c = 10$ per cent, in reasonable agreement with experiment. Since the solute atom represents an energy barrier of magnitude $\simeq bK_{max}$ to dislocation motion, it may be anticipated from the analysis of section 10.2 that the flow stress will decrease with increasing temperature above 0 K. This is found to be so experimentally, the thermal component of the flow stress being given by equation (10.10) with $\Delta F \simeq bK_{max}$, $p \simeq 2/3$ and $q \simeq 3/2$.

The analysis leading to equation (10.52) neglects several features known to exist in real alloys. First, real obstacles have a finite range of interaction with a dislocation and characteristic force–distance profiles. These details have been incorporated in several theories and some suggest a $c^{2/3}$, rather than $c^{1/2}$, dependence of the flow stress, and for many alloys a more complicated dependence probably exists. Second, more than one type of obstacle is frequently present. If τ_1 is the flow stress for one species and τ_2 that for a second, the flow stress when both are present has been found by computer simulation to be well fitted by

$$\tau = (\tau_1^2 + \tau_2^2)^{1/2} \qquad (10.53)$$

Precipitate Strengthening

When precipitates nucleate and grow, they intersect slip planes in a random fashion. A gliding dislocation must either cut through the precipitates or penetrate the array by bowing between the obstacles. It will adopt the mechanism offering the *lowest* resistance. The stress corresponding to this short-range mechanism must be compared with that required to overcome the internal stress τ_i which coherent precipitates produce in the surrounding lattice. The flow stress for the alloy is then the *larger* of the two.

Consider first the flow stress for *diffuse* forces governed by τ_i. With the assumption that the precipitates are spherical, τ_i is given by equation (10.49), where c now represents the volume fraction of precipitate. If each precipitate contains N atoms, equation (10.48) for the mean obstacle spacing is modified to

$$\Lambda = b(N/c)^{1/3} \qquad (10.54)$$

In the early stages of aging, when Λ and N are small, the weak-force conditions of section (10.6) apply and from equation (10.42) the flow stress is

$$\tau_{\text{diff}}(\text{weak}) \simeq 0.4 N^{1/9} G|\delta|^{4/3} c^{11/9} [\ln(1/c)]^{4/3} \qquad (10.55)$$

In the intermediate stages of aging, Λ and N increase until the dislocation can bend around individual precipitates, i.e. $R \simeq \Lambda$ (Fig. 10.19(a)). Then, from equations (10.39) and (10.49), the flow stress is

$$\tau_{\text{diff}}(\text{strong}) \simeq \frac{2}{\pi} G|\delta|c \ln(1/c) \qquad (10.56)$$

When overaging occurs, precipitate coherency is lost, and τ_i falls to zero and makes no contribution to the flow stress.

The *localized* forces offer two distinct resistances, as discussed above. A dislocation cutting through a precipitate experiences a resistance force K_{max} (Fig. 10.20(a)), and in the early stages of aging equation (10.46) is appropriate. The Friedel spacing Λ for precipitates is found by noting that the area

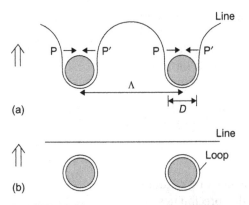

FIGURE 10.24

The Orowan mechanism,
(a) The line bows between
the obstacles until the
segments at P and P' are
parallel ($\phi_c = 0$). They then
attract and meet. (b) The
line moves forward leaving
Orowan loops around the
obstacles.

fraction of precipitate on a plane equals the volume fraction c. Since the average area of obstacle on a plane intersecting precipitates of volume Nb^3 randomly is $(\pi/6)^{1/3}(Nb^3)^{2/3}$, then

$$c = (\pi/6)^{1/3}(Nb^3)^{2/3}/\Lambda^2$$

i.e.

$$\Lambda \simeq bN^{1/3}/c^{1/2} \tag{10.57}$$

Various factors can determine $K_{\max}$. For example, (a) the precipitate may be ordered and the dislocation has to create an antiphase boundary (section 6.5), (b) the Peierls stress may be high in the precipitate, and (c) the step of height b the cutting dislocation creates around the obstacle periphery may have a high energy. The first two give $K_{\max}$ proportional to the obstacle diameter $bN^{1/3}$ and typical values are of the order of $Gb^2N^{1/3}/100$. Thus, the critical cusp angle ϕ_c is approximately $\cos^{-1}(N^{1/3}/100)$, and weak-force conditions apply only when $N < 10^5$. The flow stress from equations (10.46) and (10.57) is then

$$\tau_{\text{loc}}(\text{weak}) \simeq \frac{1}{1000}N^{1/6}Gc^{1/2} \tag{10.58}$$

As aging proceeds, N and $K_{\max}$ increase further and dislocations find it easier to bow between the obstacles than to pass through them, as shown schematically in Fig. 10.24. Under these strong-force conditions when $K_{\max} \simeq Gb^2$ and $\cos\phi_c \simeq 1$, the flow stress is obtained from equations (10.47) and (10.57). It is known as the *Orowan stress* and is

$$\tau(\text{Orowan}) \simeq 0.84\,Gb/\Lambda = 0.84N^{-1/3}Gc^{1/2} \tag{10.59}$$

In the overaged state, dislocations cannot penetrate the incoherent interface and equation (10.59) still applies.

It is now possible to explain the variation of the flow stress throughout the aging process (Fig. 10.23). In the underaged condition, dislocations cut through the precipitates, the flow stress increasing with N when either long-range internal stress (equation (10.55)) or short-range local force (equation (10.58)) determines the strength. The two effects are comparable in size and which is the dominant one depends on parameters such as $|\delta|$ and antiphase boundary energy for the alloy in question. When the precipitates are large enough for strong-force conditions to apply ($N \gtrsim 10^5$), the flow stress given by equation (10.56) is generally larger than the Orowan stress (equation (10.59)). Thus, unless coherency is lost at an early stage, further aging results in a flow stress controlled by large coherency strains (equation (10.56)). It is independent of N and corresponds to the optimum heat-treatment. In this state, a dislocation must first overcome τ_i before cutting a precipitate. When coherency is lost by overaging, a dislocation moves through the obstacles by bowing between them. The flow stress (equation (10.59)) then decreases with increasing N. Except in the very earliest stages of aging, the energy barriers offered to dislocation motion by precipitates are generally too large in comparison with kT for thermal activation to be significant, i.e. $\Delta F \gtrsim Gb^3$ in equation (10.10). The flow stress is almost athermal.

It can be seen from Fig. 10.24 that when the spacing of incoherent precipitates is not large in comparison with their size, a correction must be effected in equation (10.59) by replacing Λ by $(\Lambda - D)$. It is also seen that a dislocation moving through obstacles by the Orowan mechanism leaves a ring of dislocation around each one. These *Orowan loops* may affect the stress for subsequent dislocations by reducing the effective spacing $(\Lambda - D)$. The nature of Orowan loops and the ways in which they are affected by cross slip are discussed in section 7.5 (Fig. 7.12).

The Orowan stress is also the flow stress for *dispersion strengthening*, a process in which incoherent obstacles such as oxide particles are deliberately added to a softer metal matrix. Examples include thoria-dispersed (TD) nickel (Ni–ThO_2) and sintered aluminum powder (SAP) (Al–Al_2O_3).

The critical stress and dislocation shape characteristic of the Orowan mechanism can also arise for obstacles that can be cut by dislocations. Figure 2.18(b) shows a dislocation as it breaks away at the critical applied shear stress from a spherical void in a periodic row of voids in iron at 0 and 300 K. These images obtained by atomic-scale computer simulation (section 2.4) show atoms in the dislocation core and on the void surface. The slip plane $(1\bar{1}0)$ is the plane of the image and the Burgers vector b is $\frac{1}{2}[111]$. The dislocation was initially a straight edge dislocation lying parallel to the $[11\bar{2}]$ direction and it is seen that the segments pinned at the void surface are pulled into screw orientation in the critical condition at 0 K. The strong obstacle resistance in this case arises from two effects: (a) the dislocation strain and core energy are reduced when it penetrates a void, and (b) the dislocation has to create a surface step equal to b as it leaves the void. The obstacle strength of the void is seen to decrease with increasing temperature.

10.8 WORK HARDENING

The mechanisms discussed in the preceding sections of this chapter describe the glide of dislocations in single crystals where the resistance encountered does not change with increasing plastic strain. They neglect the increase in glide resistance that occurs when dislocations move, interact and change their distribution and density. This results in *work hardening*, which can often be exploited to advantage by raising the strength of solids shaped by plastic deformation. Taylor recognized in the earliest days of dislocation theory that work-hardening results from the interaction of dislocations. It was postulated that the flow stress τ is just the stress required to force two edge dislocations on parallel slip planes of spacing l past each other against their elastic interaction. Hence, from Fig. 4.14,

$$\tau = \alpha Gb/l \tag{10.60}$$

where α is a constant or order 0.1. The line density ρ of a network of dislocations of spacing l is $\rho \simeq l^{-2}$. With the assumption that each dislocation moves a distance x before being stopped by a network of other dislocations, it is seen from equations (3.12) and (10.60) that

$$\tau = \alpha G \left(\frac{b}{x}\right)^{1/2} \varepsilon^n \tag{10.61}$$

where n is 0.5 in the Taylor theory. Although such a parabolic relation between stress and strain is a feature of the plastic deformation of many polycrystalline materials, this is not the case for single crystals, for which equation (10.61) is derived. Also, it is known that dislocations from sources do not move as isolated defects but as groups creating slip bands. Furthermore, from the form of the stress−strain curve, it is known that barriers to dislocation motion are actually created by deformation. The incorporation of these aspects into a theory which can explain how the dislocation arrangement changes with increasing plastic strain and predict the flow stress for a given dislocation state has proved difficult, for many effects are involved.

It is now established that single crystals of a variety of metallic and non-metallic crystal structures exhibit *three-stage* behavior: the stress−strain curve for a face-centered cubic metal, for example, is shown schematically in Fig. 10.1(d) and a real example is reproduced in Fig. 10.17(a). The extent of each stage is a function of the crystallographic orientation of the tensile axis, the purity of the sample and the temperature of the test. In the body-centered cubic and prism-slip hexagonal metals, similar behavior (but at higher stress) is observed at intermediate temperatures, and in the basal-slip hexagonal metals stage I may extend almost to fracture. In all cases, the rate of work-hardening (i.e. the slope θ of the τ versus ε curve in Fig. 10.1(d)) is low in *stage I*, typically $\theta_I \simeq 10^{-4}G$, where G is the shear modulus. This linear region (often called *easy glide*) is followed by another, *stage II*, in which the slope θ is much higher, typically $\theta_{II} \simeq (3-10)\theta_I$. In *stage III*, *dynamic recovery* occurs, the τ versus ε curve is parabolic and θ decreases with increasing

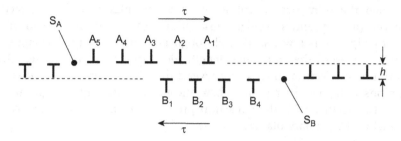

FIGURE 10.25

Dislocation sources S_A and S_B operate under resolved shear stress τ, and dislocations A_1, A_2, A_3 ... interact with B_1, B_2, B_2 ... of opposite sign to form an array of dipoles known as a *multipole*.

strain. The onset of stage III is promoted by an increase in temperature. Numerous investigations using transmission electron microscopy and measurement of surface slip steps have shown how dislocation behavior contributes to these stages.

The plastic strain in *stage I* results from dislocations moving on the slip system with the highest resolved shear stress. On this *primary system*, dislocation sources operate at the critical resolved shear stress and dislocations move over large distances (typically between $\sim 100\,\mu m$ and the crystal diameter). Many tens of dislocations contribute to slip lines observed on the surface. Electron microscopy of specimens deformed only in stage I reveals edge dislocation dipoles (see section 4.6). They are formed by the elastic interaction of dislocations of opposite sign moving in opposite directions from sources on different (but parallel) slip planes, as shown schematically in Fig. 10.25. Screw dislocations are seldom seen, because screws of opposite sign attract each other (section 4.6) and can annihilate by cross slip. The exception to this occurs in crystals of low stacking-fault energy, for then the constriction required for cross slip is difficult. Also, in overaged or dispersion-strengthened alloys (section 10.7), prismatic loops generated by cross slip (section 7.5) interact with screw dislocations to form helices. An example of helix formation is shown in Fig 3.23.

Very few dislocations are formed on *secondary* (i.e. non-primary) slip systems in stage I, but in *stage II* there is a considerable increase in multiplication and movement of secondary dislocations. The onset of this secondary activity brings stage I to an end. It occurs when the applied load (aided by internal stress from primary dislocations) and crystal orientation are such that the critical resolved shear stress on the secondary system is reached. Thus, the extent of stage I depends on crystal structure and orientation. The densities of primary and secondary dislocations are approximately equal in stage II, although the primaries contribute most to the plastic strain. Interactions between primary and secondary dislocations lead to the formation of barriers to slip (see below), so that although new primary slip lines appear as strain increases, the slip line length decreases with increasing strain. Transmission electron microscopy reveals that the density of secondary dislocations is highest in regions where the primary density is high, and the two sets, together with their reaction products, form sheets of dislocations parallel to the primary planes.

A variety of models has been proposed to explain the value of the flow stress τ in stage II for a particular dislocation distribution. To this must be added the lattice and alloying contributions discussed in sections 10.3 and 10.7. In conformity with equation (10.9), τ contains a part τ_G, which is almost temperature-independent, and a part τ^*, which is assisted by thermal activation and is temperature-dependent. τ_G arises from mechanisms that are dominated by the elastic stress field and energy of dislocations, and is therefore proportional to the shear modulus G. It has the form

$$\tau_G \propto Gb/l \qquad (10.62)$$

where length l depends on the controlling mechanism. For example, if τ_G is the stress required to make parallel dislocations of opposite sign pass on parallel slip planes of spacing h (Fig. 10.25), as in stage I, then $l = h$, as seen by putting $y = h$ in Fig. 4.14.

When forest dislocations of spacing l_f intersecting the slip plane provide obstacles to glide, relation (10.62) applies with $l = l_f$, and if the forest arrangement is random, then l_f^{-2} is approximately equal to the density ρ_f of those dislocations, i.e.

$$\tau \propto Gb\sqrt{\rho_f} \qquad (10.63)$$

Three contributions to τ arise. The first is τ_G and is the stress required to overcome attractive junctions, as discussed below. The other two contribute to τ^* and are short-range in nature. One is the applied stress required to create a jog by forest intersection (section 7.7) and the other the stress necessary to move a jogged dislocation by vacancy generation (section 7.3). The contribution due to jog creation arises from the energy to form a single jog E_j. In the absence of thermal activation, this is equal to the work done by the external load when a dislocation segment of length l_f under force τb per unit length creates a jog and moves forward by b; i.e. $E_j = \tau b^2 l_f$. Treating the jog as a dislocation element of length b, E_j is approximately $0.2Gb^3$ (section 7.2). The stress at $T = 0$ K is therefore given by

$$\tau^*(0) = 0.2\ Gb/l_f \qquad (10.64)$$

The stress to create a vacancy by jog climb is given by equation (7.1). Taking $E_f^v \simeq 8kT_m$ (section 1.3), which is typically about $0.1Gb^3$, the stress becomes

$$\tau^*(0) = 0.1\ Gb/l_f \qquad (10.65)$$

Thus, both contributions to τ^* at $T = 0$ K have the form

$$\tau^*(0) = \alpha Gb\sqrt{\rho_f} \qquad (10.66)$$

where α is approximately 0.1 to 0.2. Thermal activation decreases τ^*, i.e. α, with increasing temperature. In general, τ_G dominates and τ^* makes only a weak contribution to τ at low homologous temperatures. Exceptions to this include alloys and body-centered cubic crystals, in which thermally-activated solid-solution and Peierls-stress effects can lead to τ^* being larger than τ_G at low to moderate temperatures.

The mechanisms giving rise to the dominant τ_G contribution to τ comes from attractive interactions between dislocations on different slip systems. The collinear interaction and reactions that form the Lomer–Cottrell lock, the Hirth lock and the glissile junction in the face-centered cubic metals were described in sections 5.6 and 7.6. Their contributions to the critical stress for slip on any one system can be combined in the form

$$\tau^i = Gb\sqrt{\sum_j a^{ij}\rho^j} \qquad (10.67)$$

where τ^i is the stress for slip system i, ρ^j the dislocation density in slip system j and coefficient a^{ij} is a measure of the strength of interaction between systems i and j. Kubin and co-workers have obtained the magnitudes of the strength coefficients for the possible interactions by dislocation dynamics computer simulation (section 2.4) (see Further Reading). The collinear interaction (section 7.6) makes by far the strongest contribution ($\sqrt{a^{ij}} \simeq 0.8$ compared with 0.2–0.4 for the others). Further understanding of strain hardening should be gained by using equation (10.67) in models in which the dislocation density terms change with stress and strain in a manner compatible with experiment.

The equations above do not give a work-hardening rate θ, nor do they predict the form of the dislocation arrangement. There is, as yet, no completely statisfactory theory of stage I. Most models have considered the applied stress τ_G for one dislocation to pass other, parallel dislocations in dipole/multipole configurations, which, as explained above, are characteristic of this stage. Unfortunately, multipoles are probably very mobile unless they contain exactly equal numbers of dislocations of opposite sign, so that it is necessary to invoke some (unexplained) trapping mechanism for multipoles to predict the observed θ. It is possible that forest dislocations play a role here. In stage I, the long-range internal stress field from dislocations in dipole pairs is weak and θ correspondingly small. In stage II, secondary slip leads to strong obstacles and pile-ups of primary dislocations. The large internal stress that results from these leads to further secondary slip and the formation of stable dislocation boundaries with long-range stress fields. New dislocations are generated in relatively soft regions of the crystal, before being blocked themselves. The slip line length therefore decreases with increasing strain and τ_G increases rapidly. Again, however, there is no completely satisfactory theory which accounts quantitatively for the net contributions from the forest and primary dislocations.

Stage II ceases and *stage III* begins when dislocations leave their original slip planes. Except for high temperatures at which edge dislocations can climb, this occurs by the cross slip of screw dislocations, as confirmed by cross-slip traces seen by surface examination. The attraction and annihilation of screws of opposite sign leads to a reduction in dislocation density (and internal stress) in stage III, and the edge dislocations rearrange to form low-angle boundaries. This change of structure is promoted by the applied stress and is thus known as dynamic recovery. Since cross slip is hindered by low stacking-fault energy (section 5.3) and assisted by thermal vibrations, stage III dominates the stress–strain curve for high stacking-fault energy crystals at moderate to high temperatures, as illustrated by the dashed curve in Fig. 10.1(d). The stress at which stage III starts decreases exponentially with increasing temperature and stacking-fault energy.

It is clear that analytical theory alone has not explained all the features of work hardening and that answers to some questions are coming from computer simulation of the behavior of dislocation arrangements that mimic those in real crystals. In addition to the detailed studies of junction reactions such as those referred to above, the motion and interaction of many dislocations can be modeled in the continuum approximation by the method of dislocation dynamics (section 2.4). An example of the change in dislocation arrangement found by simulation of a molybdenum single crystal $10 \times 10 \times 10\ \mu m^3$ was shown in Fig. 2.20. The corresponding curves for applied stress and dislocation density versus strain are plotted in Fig. 10.26. The stress was calculated from Young's modulus for molybdenum times the elastic strain, and the latter was obtained by subtracting the plastic strain, computed from the motion of the dislocation segments (equation (3.12)), from the total strain. The stress–strain plot has the form described earlier for a ductile metal, i.e. the deformation is mainly elastic for small strains and plastic after yielding. The model exhibits work hardening with a rate $(d\tau/d\varepsilon)$ of approximately 6 GPa. The yield drop seen in Fig. 10.26(a) reflects the choice for the initial arrangement of dislocations aligned preferentially along

FIGURE 10.26

(a) Stress versus strain curve and (b) dislocation density versus strain for the dislocation dynamics simulation of single crystal molybdenum described in section 2.4 (Fig. 2.20). *(From Bulatov and Cai (2006),* Computer Simulations of Dislocations, *Oxford University Press. With permission from Oxford University Press (www.oup.co.uk).)*

screw orientations and a dislocation mobility law with a high drag coefficient for screws. As explained in section 10.5, a stress drop is associated with a rapid increase in density of mobile dislocations at yield.

10.9 DEFORMATION OF POLYCRYSTALS

There are important differences between plastic deformation in single-crystal and polycrystalline materials. In the latter, the individual crystals have different orientations, and the applied resolved shear stress for slip (equation (3.1)) varies from grain to grain. A few grains yield first, followed progressively by the others. The grain boundaries, being regions of considerable atomic misfit, act as strong barriers to dislocation motion, so that unless the average grain size is large, stage I easy-glide exhibited by single crystals does not occur in polycrystals. The stress—strain curve is therefore not simply a single-crystal curve averaged over random orientations. Furthermore, the internal stresses around piled-up groups of dislocations at the boundaries of grains that have yielded may cause dislocation sources in neighboring grains to operate. Thus, the macroscopic yield stress at which all grains yield depends on grain size. Finally, a grain in a polycrystal is not free to deform plastically as though it were a single crystal, for it must remain in contact with, and accommodate the shape changes of, its neighbors. An inability to meet this condition leads to a small strain to failure, as typified by Fig. 10.1(b).

Considering the last point first, it is seen from section 4.2 that when volume remains constant, as it does during plastic deformation of crystals, a general elastic shape change is fully specified by five independent components of strain. Extending this to the ability to undergo general *plastic* shape changes, *five independent slip systems* must be able to operate, a result known as the *von Mises condition*. An independent system produces a shape change which cannot be obtained by combinations of other systems. In the face-centered cubic metals, the twelve $\langle 1\bar{1}0\rangle\{111\}$ slip systems provide five independent systems and satisfy the condition. So, too, do the slip systems in body-centered cubic metals. In other structures (see Chapter 6) there are often insufficient systems, except at high temperature. For example, hexagonal metals in which basal slip is strongly preferred have only three slip systems of which two are independent — neglecting any contribution from deformation twinning — and show little ductility at low temperatures.

Equation (3.1) shows that the tensile yield stress σ_y and the critical resolved shear stress τ_c for a single crystal which slips on one plane in one direction are related by the Schmid factor $(\cos \phi \cos \lambda)$. It may be rewritten

$$\sigma_y = m\tau_c \tag{10.68}$$

For polycrystals in which the grains have random orientations, this may be generalized to

$$\sigma_y = M\tau_c \tag{10.69}$$

where M is known as the *Taylor factor*. By averaging the stress over the grains and considering the most-favored slip systems, it has been shown that $M \simeq 3$ for the face-centered and body-centered cubic metals.

From experimental measurement of the yield stress of polycrystalline aggregates in which grain size d is the only material variable, it has been found that the *Hall–Petch relationship* is satisfied:

$$\sigma_y = \sigma_0 + k_y d^{-n} \tag{10.70}$$

where exponent n is approximately 0.5, k_y is a material constant and σ_0 is a constant stress of uncertain origin. The flow stress beyond yield follows a similar form. Hence, at cold-working temperatures, where grain boundaries do not contribute to creep, high yield and flow stresses are favored by small grain size. One rationalization of equation (10.70) is that a pile-up at a grain boundary in one grain can generate sufficiently large stresses to operate dislocation sources in an adjacent grain at the yield stress. For example, the stress τ_1 on the leading dislocation generated by source S_1 in grain 1 under resolved shear stress τ (Fig. 10.27) is seen from equations (9.27) and (9.28) to be

$$\tau_1 = \frac{d}{A}\tau^2 \tag{10.71}$$

where $A \simeq Gb/2\pi$. If it is assumed that source S_2 in grain 2 operates when τ_2 reaches a critical value τ_1^*, then

$$\sigma_y = k_y d^{-1/2} \qquad (10.72)$$

where k_y is $m(A\tau_1^*)^{1/2}$, and m converts resolved shear stress to tensile stress. The Hall–Petch relationship follows by assuming the dislocations from S_2 encounter a friction stress or internal back-stress σ_0. Although such a pile-up model was first used to explain the relationship, other interpretations, such as that based on grain-boundary dislocations (Li and Chou, 1970), are possible.

FIGURE 10.27
Schematic illustration of a pile-up formed in grain 1 under an applied resolved shear stress τ. S_2 is a source in grain 2. The trace of the preferred slip plane in each grain is marked by a dashed line.

Equation (10.70) predicts that high values of yield stress can be obtained if metals can be processed to have d less than about 100 nm. Although experiments on such *nanograined* materials have shown that σ_y values exceeding 1 GPa can be achieved, the Hall–Petch relationship breaks down in many cases, i.e. σ_y no longer increases with decreasing d and may actually decrease. This is believed to be due to the fact that the stress to operate Frank–Read sources within nanograins would be very high and could not be reached before other processes result in deformation. These might involve grain boundary sliding, grain boundary migration or dislocation generation in boundaries. There is as yet no comprehensive model of plasticity of nano-grain materials (see Further Reading).

10.10 DISLOCATIONS AND FRACTURE

Surface irregularities and cracks are particularly potent sources of disloca-tions in solids under stress. Indeed, the generation of dislocations in the vicinity of crack tips plays a vital role in determining whether material exhi-bits brittle or ductile behavior when fracture occurs, i.e. whether crack propa-gation is preceded by significant dislocation emission at the crack tip or not. In ductile metals, yielding at the tip of a crack can blunt it and crack propa-gation is accompanied by extensive plastic flow. However, many materials, e.g. ceramics, semiconductors and refractory metals, exhibit a *brittle-to-ductile transition* as their temperature is raised. Some, such as silicon, show a very sharp transition over a narrow temperature range, but most do not and their transition is soft, occurring over a wide temperature range of typically 50–100°. Dislocations influence the behavior by affecting the stress field near the crack tip before cleavage occurs. Experiments show that dislocations glide and multiply near the crack tip over the temperature range of the transition. A soft transition is assisted by a high density of available dislocation sources along the crack front.

To a first approximation, if material under an applied tensile stress σ contains a crack of length c perpendicular to the stress axis, the stress field near the tip of the crack is proportional to $\sigma\sqrt{c}$ and is magnified strongly in the tip region. For sharp cracks it is convenient to use the *stress intensity factor* K to describe the stress around the crack:

$$K = \alpha\sigma\sqrt{\pi c} \qquad (10.73)$$

where α is a constant approximately equal to 1. The stress field near the crack tip varies as $K/\sqrt{r}$, where r is the distance from the tip. Local fracture occurs when K reaches a critical value K_{frac} corresponding to there being sufficient strain energy in the crack tip stress field to create new fracture surfaces. For a given crack length, the fracture stress and K_{frac} are given by equation (10.73).

Dislocation activity near the tip of a crack can *shield* the tip from the full effects of the applied stress and result in a local stress intensity factor at the crack tip, K_{tip}, that is lower than that given by equation (10.73), i.e.

$$K_{tip} = K - K_{disl} \qquad (10.74)$$

where K_{disl} arises from crack-tip shielding. With no dislocation activity to shield the crack, as in ideally brittle materials, fracture occurs when K_{tip} equals the critical value K_c as given by the Griffith theory for pure cleavage, i.e. $K_{frac} = K_c$. With limited dislocation activity, fracture can still occur but only when $K_{frac} > K_c$. When extensive dislocation occurs, K_{disl} may be sufficiently large that K_{tip} given by equation (10.74) does not reach the value for fracture and the material will be ductile.

The length scale of the shielding process is of the order of μm and too large for simulation by atomic-scale computer modeling. Although this method has been applied to study dislocation nucleation at the crack tip itself, e.g. at ledges, evidence points to the difficulty of actually creating sufficient dislocations in this way to provide significant shielding. The models that have been developed to explain the role of dislocations in the transition from brittle to ductile behavior assume that the dislocation sources near the crack tip pre-exist. The models employ elasticity theory to calculate the stress field of many moving dislocations. The simplest, i.e. two-dimensional, approach to model the effect of dislocations emitted from a source near a crack tip under an applied stress is illustrated schematically in Fig. 10.28. The first dislocation labeled 1 is emitted by the source when K is well below K_c and will be able to move away from the crack tip if the force it experiences due to the crack stress field is sufficient: this should include, for example, the image stress component (section 4.8) due to the crack surface. The sign of dislocation 1 is such that its stress field reduces the stress at the source and so the applied K will have to be increased to allow dislocation 2 to be emitted and glide away. Dislocation 2 adds to the back stress at the source and K will have to increase again for dislocation 3 to be emitted, and so on for subsequent dislocations. As a result of the shielding effect of the emitted dislocations, the applied stress intensity K at fracture will be greater than K_c.

FIGURE 10.28
Schematic illustration of the generation of shielding dislocations from a source just ahead of a crack.

The dynamics of the situation depend on the rate of increase of applied load and the time the shielding dislocations remain in the vicinity of the crack before gliding away. The latter can be studied in the framework of dislocation dynamics by simulating dislocation movement using a realistic dependence of dislocation velocity on the local stress and temperature. Two- and three-dimensional models of this sort provide a good description of the rise in K during the fracture transition of simple materials. A comparison of model and experimental results for single crystal tungsten is plotted in Fig. 10.29 for five different values of the applied strain rate.

FIGURE 10.29
Fracture toughness versus temperature obtained by experiment and modeling for single crystal tungsten at five applied strain rates. The vertical dotted lines indicate the experimental transition temperatures. *(From Fig. 4 of Tarleton and Roberts, Phil. Mag., 89, 2759, 2009. With permission from Taylor and Francis Ltd (http://www.informaworld.com).)*

FURTHER READING

Argon A: *Strengthening mechanisms in crystal plasticity*, 2007, Oxford University Press.

Ashby MF, Jones DRH: *Engineering materials*, ed 3, vols 1 and 2, 2005, Butterworth-Heinemann.

Bacon DJ, Osetsky YN, Rodney D: Dislocation–obstacle interactions at the atomic level. In Hirth JP, Kubin L, editors: *Dislocations in solids*, vol 15, 2009, North-Holland, p. 1.

Basinski SJ, Basinski ZS: Plastic deformation and work hardening. In Nabarro FRN, editor: *Dislocations in solids*, vol 4, 1979, North-Holland, p. 261.

Brown LM, Ham RK: Dislocation-particle interactions. In Kelly A, Nicholson RB, editors: *Strengthening methods in crystals*, 1971, Elsevier, p. 12.

Cherkaoui M, Capolungo L: *Atomistic and continuum modeling of nanocrystalline materials: Deformation mechanisms and scale transition*, 2009, Springer.

Christian JW: *The theory of transformations in metals and alloys (part I and II)*, ed 3, 2002, Pergamon.

Cottrell AH, Bilby BA: Dislocation theory of yielding and strain ageing in iron, *Proc Phys Soc* A62:49, 1949.

Edington JW, Melton KN, Cutler CP: Superplasticity, *Prog Mater Sci* 21:61, 1976.

Eshelby JD: The continuum theory of lattice defects, *Solid State Phys* 3:79, 1956.

Eshelby JD: Determination of the elastic field of an ellipsoidal inclusion, *Proc Roy Soc* A241:376, 1957.

Friedel J: *Dislocations*, 1964, Pergamon.

Gerold V: Precipitation hardening. In Nabarro FRN, editor: *Dislocations in solids*, vol 4, 1979, North-Holland, p. 219.

Gifkins RC, editor: *Strength of metals and alloys*, 3 vols, 1983, Pergamon.

Haasen P: Solution hardening in fcc. metals. In Nabarro FRN, editor: *Dislocations in solids*, vol 4, 1979, North-Holland, p. 155.

Hertzberg RW: *Deformation and fracture mechanics of engineering materials*, 1996, John Wiley.

Hirsch PB: Work hardening. In Hirsch PB, editor: *The physics of metals: 2 defects*, 1975, Cambridge University Press, p. 189.

Hirth JP, Lothe J: *Theory of dislocations*, 1982, Wiley.

Johnston WG: Yield points and delay times in single crystals, *J Appl Phys* 33:2716, 1962.

Kelly A, Groves GW, Kidd P: *Crystallography and crystal defects*, 2000, John Wiley.

Kocks UF, Argon AS, Ashby MF: Thermodynamics and kinetics of slip, *Prog Mater Sci* 19:1, 1975.

Kubin L, Devincre B, Hoc T: The deformation stage II of face-centered cubic crystals: Fifty years of investigations, *Int J Mat Res* 100:10, 2009.

Li JCM, Chou YT: The role of dislocations in the flow stress grain size relationship, *Met Trans* 1:1145, 1970.

Martin JW: *Precipitation hardening*, ed 2, 1998, Butterworth-Heinemann.

Nabarro FRN: *The theory of crystal dislocations*, 1967, Oxford University Press.

Nabarro FRN: Solution and precipitation hardening. In Hirsh PB, editor: *The physics of metals: 2 defects*, 1975, Cambridge University Press, p. 152.

Nabarro FRN, Duesbery MS: *Various Chapters in 'Work Hardening'*, *Dislocations in solids*, vol 11, 2002, North-Holland.

Nowick AS, Berry BS: *Anelastic relaxation in crystalline solids*, 1972, Academic Press.

Pande CS, Cooper KP: Nanomechanics of Hall—Petch relationship in nanocrystalline materials, *Prog Mater Sci* 54:689, 2009.

Ritchie IG, Fantozzi G: Internal friction due to the intrinsic properties of dislocations. In Nabarro FRN, editor: *Dislocations in solids*, vol 9, 1992, North-Holland. chap. 45

Saada G, Dirras G: Mechanical properties of nanograined metallic polycrystals. In Hirth JP, Kubin L, editors: *Dislocations in solids*, vol 15, 2009, North-Holland, p. 199.

Smith E: Dislocations and cracks. In Nabarro FRN, editor: *Dislocations in solids*, vol 4, 1979, North-Holland, p. 363.

Van Swygenhoven H, Derlet PM: Atomistic simulations of dislocations in FCC metallic nano-crystalline materials. In Hirth JP, editor: *Dislocations in solids*, vol 14, 2008, North-Holland, p. 1.

Vitek V, Paidar V: Non-planar dislocation cores: A ubiquitous phenomenon affecting mechanical properties of crystalline materials. In Hirth JP, editor: *Dislocations in solids*, vol 14, 2008, North-Holland, p. 439.

Was GS: *Fundamentals of Radiation Materials Science*, 2007, Springer.

Xu G: Dislocation nucleation from crack tips and brittle to ductile transitions in cleavage fracture. In Nabarro FRN, Hirth JP, editors: *Dislocations in solids*, vol 12, 2004, North-Holland, p. 81.

The SI System of Units

Base-units

metre (m) — length
kilogram (kg) — mass
second (s) — time
ampere (A) — electric current
kelvin (K) — thermodynamic temperature
candela (cd) — luminous intensity

Some derived units

	Name of SI unit	SI base-units
frequency	hertz (Hz)	$1\,Hz = 1\,s^{-1}$
force	newton (N)	$1\,N = 1\,kg\,m\,s^{-2}$
work, energy	joule (J)	$1\,J = 1\,N\,m$

Multiplication factors

	Prefix	Symbol
10^{12}	tera	T
10^{9}	giga	G
10^{6}	mega	M
10^{3}	kilo	k
10^{-3}	milli	m
10^{-6}	micro	μ
10^{-9}	nano	n
10^{-12}	pico	p
10^{-15}	femto	f
10^{-18}	atto	a

Some useful conversions

1 angstrom $= 10^{-10}\,m = 100\,pm$ or $0.1\,nm$
1 kgf $= 9.8067\,N$

1 atm	$= 101.33 \text{ kN m}^{-2}$
1 torr	$= 133.32 \text{ N m}^{-2}$
1 bar	$= 0.1 \text{ MN m}^{-2}$
1 dyne	$= 10^{-5} \text{ N}$
1 dyne cm^{-1}	$= 1 \text{ mN m}^{-1}$
1 dyne cm^{-2}	$= 0.1 \text{ N m}^{-2}$
1 cal	$= 4.1868 \text{ J}$
1 erg	$= 10^{-7} \text{ J} = 0.1 \text{ } \mu\text{J}$
1 ergcm^{-2}	$= 1 \text{ mJ m}^{-2} = 1 \text{ mN m}^{-1}$
1 inch	$= 25.4 \text{ mm}$
1 lb	$= 0.4536 \text{ kg}$
1 lbf	$= 4.4482 \text{ N}$
1 lbf in^{-2}	$= 6.895 \text{ kN m}^{-2}$
1 lb in^{-1}	$= 175 \text{ N m}^{-1}$
1 eV	$= 1.6 \times 10^{-19} \text{ J} = 0.16 \text{ aJ}$

Index

A

Ab initio calculations, 34—5, 112, 114, 121
Activation energy, 208
Activation volume, 208
Admissible defects, 186—9
Aging, 222, 233—7
Alloy strength, 232—7
Anion vacancies, 10
Antiphase boundaries (APB), 127, 128
Asymmetry of slip, 119
Athermal activation, 209
Atmosphere, Cottrell, 220
Atom probe tomography, 33—4
Atomic planes, 15
Atomic-scale simulation, 21, 34—8, 119—20
Attractive junctions, 149—53

B

Bardeen—Herring source, 167—8
Basal planes, 109—10, 115, 117
Basal slip, 112—14
bcc *see* Body-centered cubic metals
Bicrystals, 187—9, 200
Body-centered cubic metals (bcc), 5, 118—24
   octahedral interstitials, 219
   twinning, 196, 197
   work hardening, 238
Bond reconstruction in the core, 132—3
Boundaries
   dislocations, 175—7
   movement of, 193—202
Bragg's law, 22—3
Bright field images, 23
Brittle materials, 205

Brittle-to-ductile transitions, 245
Bubble models, 12
Bulk modulus, 66
Burgers circuit, 16—19, 87
Burgers vector, 16—19, 26, 94

C

Cation vacancies, 10
Channeling of dislocations, 228
Chemical forces, 80
Chemical interaction, 220
Circuit mapping, 189—91
Circular edge loops, 98
Circular shear loops, 98
Cleavage, 246
Climb, 53—6
   conservative, 59
   experimental observation, 56—9
   forces, 79—81
   jogs, 241
   multiplication by, 167—8
   rate of, 80—1
Close-packed hexagonal structures
   see Hexagonal metals
Coherent interfaces, 191
Coincident site lattice (CSL), 188
Cold work, 171
Collinear reactions, 150, 152, 241
Column approximations, 23
Commensurate interfaces, 191
Composition planes, 12
Compression, 66
Computer simulation, 21—2, 34—41, 119—20, 199, 214
Conservative climb, 59
Conservative motion, 43, 195—9
Constriction of extended dislocation, 91—3
Continuum-level simulation, 38—40

Cores
   bond reconstruction, 132—3
   disregistry, 120—1
   energy of, 72
   radius, 69
   structure of, 119—21, 210—13
Cottrell atmosphere, 220
Covalent crystals, 130—3, 220
cph *see* Hexagonal metals
Crack, 245—8
Critical resolved shear stress (CRSS), 127, 206
Critical shear stress for slip, 13—14
Cross slip, 48—50, 91—3, 106, 242
Crowdion interstitials, 123
CRSS (critical resolved shear stress), 127, 206
Crystalline solids
   boundaries, 171—204
   strength of, 205—49
Crystals
   atom positions, 1
   defects in, 8—13
   grain boundaries, 12
   growth rates, 14
   point defects, 8—10
   stacking faults, 10—11
   structure, 1—8, 20, 135—6
   twin boundaries, 12—13
   volume defects, 13
CSL (coincident site lattice), 188
Cubic crystals, 3, 4

D

D-Shockley partial, 98
Dangling bonds, 132
Dark field images, 23, 27
Debris, 144
Decoration methods, 30—1

Defects, admissible, 186–9
Deformation
    experimental observation, 198
    plastic, 171–5, 205
    polycrystals, 243–5
    twinning, 12–13, 122, 195–9
Degenerate cores, 120–1
Degrees of freedom, 174, 175
Diamond-cubic structures, 131
Dichromatic patterns, 188, 196
Diffraction mode, 23
Diffuse forces, 229–30, 235
Diffusion-assisted motion, interfacial
    defects, 199–201
Dilatation, 65
Dipoles, 79, 143–4, 145
Disconnections, 188
Discrete dislocation dynamics, 38–40
Dislocation(s), 13–19, 161, 162–3
    arrays, 171–204
    body-centered cubic metals,
        118–24
    boundaries, 175–7
    Burgers circuit, 16–19
    Burgers vector, 16–19
    channeling, 228
    cores, 69, 72, 119–21, 132–3,
        210–13
    covalent crystals, 130–3, 220
    definition, 16
    density, 19, 157
    dipoles, 143–4
    dissociation, 91, 103
    dynamics simulation, 21–2,
        38–40
    elastic energy, 122, 185
    elastic properties, 63–83
    energy, 116, 122, 157, 185, 213
    entropy, 157
    epitaxial interfaces, 191–3
    extended, 90, 91, 147–9
    face-centered cubic structures,
        85–107
    forces between, 75–9
    forces on, 73–5
    fracture, 245–8
    in freshly grown crystals, 158–9
    geometry of, 15–16
    glide set, 132
    hexagonal close-packed metals,
        109–18
    imperfect, 110
    interfacial, 186–93
    intersections, 137–9, 155

ionic crystals, 124–6, 219
jogs, 137–55
    in layer structures, 133–4
    locking, 99–102, 220–4
    loops, 56–7, 97–8, 122–4,
        144–7
    misfits, 158, 192–3
    mixed, 24–5, 49, 70, 72–3
    movement of, 43–62, 139–42
    multiplication of, 157–69
    network, 177
    nodes, 19, 58
    nucleation, 99, 158, 159–63, 215
    observations, 21–41
    origin of, 157–69
    partial, 85–7, 99–102, 134
    perfect, 18, 85, 94, 110, 122, 131
    pile-ups, 203–4, 245
    plastic strain, 59–61
    point defect interactions, 8–10,
        20, 216–20
    polymer crystals, 134–5
    screw, 15–16, 24, 29, 67–9,
        105–6, 113–14, 119
    sessile, 95, 96, 100, 101
    slip, 45–6
    spiral, 57–9
    strain energy, 67, 71–3, 184–6
    stress fields, 67–70, 174, 176–7,
        181–4
    superlattices, 126–30
    superpartial, 128–30
    velocity, 50–3, 210
    walls, 183–4
    see also Edge dislocations
Dispersion strengthening, 237
Displacement damage, 28
Displacement difference, 210–12
Displacive transformation, 201–2
Disregistry in the core, 120–1, 210–12
Dissociation, 91, 103, 114, 212
Double cross slip, 49, 144
Double-kink nucleation, 215
Drag coefficient, 52
Ductile materials, 205
Dynamic recovery, 238–9
Dynamic strain aging, 222

E
EAM (Embedded Atom Model), 36
Easy glide, 238
Edge dislocations, 15, 16, 69–70,
    113, 114, 124–6
    atom positions, 92, 212

Burgers vector, 17, 212
    climb, 53, 61
    disregistry, 212
    formation, 45
    intersections, 137–9
    jogs, 55, 140
    movement, 46
    negative climb, 53
    pile-ups, 203
    plastic deformation, 48
    plastic displacement, 60
    plastic strain, 61
    positive and negative climb, 53
    stable positions, 79
    stress fields, 182, 183
    velocity of, 51, 53
Effective charge, 124–6
Effective pressure, 70
Elastic energy, 122, 185
Elastic interactions, 218–19, 221
Elastic properties of dislocations,
    63–83
Elasticity theory, 63–7, 122
Electrical interactions, 219
Electron microscopy, 22–8
    dislocations, 24–5
    image simulation, 26–8, 29
    lattice imaging, 26
    limiting factors, 28
    planar defects, 25–6, 27, 28
    principles, 22–4
Elementary jogs, 125, 139–42
Embedded Atom Model (EAM), 36
Entropy of dislocations, 157
Epitaxial interfaces, 191–3
Equilibrium concentration of defects,
    9
Escaig stress, 92
Eshelby twist, 68–9
Etch-pit studies, 29–30, 31, 32, 33
Extended dislocations and jogs, 90,
    91, 147–9
Extended stacking-fault nodes,
    153–4
External stress, 63
Extrinsic point defects, 9
Extrinsic stacking faults, 11, 95, 111

F
Face-centered cubic structures (fcc),
    5–6, 85–107, 112
    boundaries, 180
    stacking faults, 11
    work hardening, 238

Field evaporation, 32
Field ion microscopy, 32−3
Flow stress, 207−10, 228−32, 235
Forces between dislocations, 75−9
Forces on dislocations, 73−5
Forest dislocations, 137, 240
Fracture, 245−8
Frank net, 19
Frank partial dislocations, 94−9, 104
Frank−Read sources, 163−6, 245
Frank's formula, 177−8, 179
Frank's rule, 73, 76, 89
Free energy, 208
Frenkel defects, 10
Friction coefficient, 52
Friedel relation, 231

G

Gamma surface, 88
General low-angle boundaries,
    177−81
General tilt boundaries, 175
Geometrical models, 15−16, 112
Geometrically necessary dislocations,
    173
Geometry of dislocations, 15−16
Glide, 43, 195−9
Glide bands, 166−7
Glide cylinder or prism, 56
Glide force, 39, 74
Glide planes, 54, 56, 195
Glide set dislocations, 132
Glide system, 126
Glissile loops, 118
Glissile reactions, 150, 151
Grain boundaries, 12, 33, 36
    energy, 185−6
    further reading, 20, 204
    interfacial defect motion, 201
    large-angle, 172
    sources, 169
Grain growth, 172
Growth rates, crystals, 14

H

Habit planes, 196, 202
Hall−Petch relationship,
    244, 245
Hard sphere model of atoms, 4,
    6−7, 10−11, 110−11
hcp see Hexagonal metals
Helical dislocations, 57−9
Hexagonal crystals, indices, 7−8

Hexagonal metals, 6, 7, 109−18,
    180−1
    slip systems, 244
    work hardening, 238
Hirsch mechanism, 146
Hirth lock, 100
Homogeneous nucleation of
    dislocations, 159−60
Hooke's law, 66

I

Image forces, 81−2
Image simulation, 26−8, 29
Imperfect dislocations, 110
Impurity atoms, 9, 10, 31
Inhomogeneity interactions, 219
Interfaces, 186−93
    defects, 189−91, 199−201
    glide, 195−9
    line defect identification, 189−91
Internal stress, 63
Interphase boundaries, 199, 201
Intersection of dislocations, 137−9,
    155
Interstitials
    defects, 8−9, 10
    jogs, 55, 140, 141−2
    loops, 97, 115−18, 123
    octahedral, 218−19
Intrinsic point defects, 8−9
Intrinsic stacking faults, 11, 88, 95,
    116
Ionic crystals, 124−6, 219
Irradiation hardening, 227

J

Jog lines, 104
Jogs, 54−5, 125, 137−55
    climb, 241
    diffusion, 56
    electrical charge, 219−20
    elementary, 125, 139−42
    energy, 240
    prismatic loops, 144−7
Junctions, 149−53

K

Kear−Wilsdorf lock, 130
Kink pair nucleation, 215
Kinks, 54, 55, 125, 214−16

L

Large-angle boundaries, 172, 175
Lattice imaging, 26

Lattice resistance, 210−16
Lattice structures, 15
Lattice translation vectors, 18
Layer structures, 133−4
Left-handed screw dislocations, 15,
    16, 47
Line defects in interfaces, 186−9
Line dislocation dynamics, 38−40
Line tension, 74−5, 80, 229−30
Linear elasticity, 63−4
Localized forces, 230−2, 235
Locking, 99−102, 220−4
Lomer−Cottrell lock, 101, 102
Lomer dislocations, 100
Lomer lock, 99, 101
Long jogs, 142−4
Long-range stress fields, 174, 176−7,
    183
Loops, 97−8, 122−4, 144−7
    formation, 162−3
    prismatic, 56−7
Low-angle boundaries, 171−5,
    177−81, 193−4
Lüders band, 206

M

Martensite, 201−2
Microplasticity, 205
'Microwhiskers', 14
Migration energy, 9
Miller−Bravais indices, 7−8, 110
Miller indices, 1−3, 4, 7
Misfit dislocations, 158, 192−3
Misfit parameter, 217
Misfit volume, 217
Mixed dislocations, 24−5, 49, 70,
    72−3
Mobile dislocations, 226−7
Molecular dynamics, 35, 36−8
Molecular statics, 35, 36
Motif units, 1
Movement of boundaries, 193−202
Movement of dislocations, 43−62,
    139−42
Multi-junctions, 153
Multiple cross glide, 166−7
Multiplication of dislocations,
    157−69
Multipoles, 239, 241−2

N

Nanograins, 245
Nanowires, 68−9
Negative climb, 54, 80

Negative edge dislocations, 16
Network of dislocations, 177
Nodes, 19, 58
    density, 39
    dimension, 154
    extended, 153–4
    forces on, 39
Non-basal slip, 114–15
Non-degenerate cores, 120–1
Normal strains, 64
Nucleation
    of dislocations, 99, 158, 159–63,
        215
    precipitates, 233

O

Observations of dislocations, 21–41
Octahedral interstitials, 218–19
Optical methods, 31
Ordered alloys, 127
Origin of dislocations, 157–69
Orowan equation, 59
Orowan mechanism, 146, 236, 237
Overageing, 235, 236, 237

P

Parallel edge dislocations, 75–8
Parallel screw dislocations, 79
Partial dislocations, 85–7, 99–102,
    134
Peierls energy, 213–14, 215
Peierls–Nabarro stress, 46, 210–16
Perfect dislocations, 18, 85, 94, 110,
    122, 131
Phonons, 52
Pile-ups, 203–4, 245
Pipe diffusion, 55–6, 59
Planar defects, 10–11, 25–6, 27, 28
Plane strain, 70
Plastic deformation, 171–5, 205
Plastic strain, 59–61
Point defects, 8–10, 20, 216–20
Poisson's ratio, 66
Polarized cores, 120–1
Polycrystalline materials, 114, 206,
    243–5
Polygonization, 172
Polymer crystals, 134–5
Portevin–Le Chatelier effect, 222
Positive climb, 53, 54
Positive edge dislocations, 15, 16, 47
Precipitates, 31, 37, 116, 117, 232–3
    nucleation, 233
    strengthening, 235–7

Primary dislocations, 239–40, 242
Prism planes, 109–10, 112, 113,
    114
Prism slip, 113, 238
Prismatic dislocation loops, 56–7,
    97–8, 116, 117, 144–7,
    161–3, 239–40
Proof stress, 205
Pyramidal planes, 110

Q

Quenching, 104

R

Radiation damage, 227
Random arrays, 228–32
Ray diagrams, 22
Read–Shockley formula, 185, 186
Reconstruction in the core, 132–3
Recovery, 171–5
Recrystallization, 171–5
Regenerative multiplication, 163–8
Repulsive junctions, 149–53
Right-handed screw dislocations, 15,
    16, 47, 59, 139
Rocksalt structure, 124

S

S-Shockley partial, 98
Schmid factor, 45
Schottky defects, 10
Screw dislocations, 15–16, 24, 29,
    67–9, 105–6, 113–14, 119
    atom positions, 47
    Burgers vector, 17
    cross slip, 242
    effective charge, 126
    face-centered cubic structures, 91
    glide, 121
    jogs, 55, 140–3, 149
    kinks, 55
    movement, 47
    plastic deformation, 48
    velocity of, 51
Secondary dislocations, 239–40, 242
Seed crystals, 158
Self-interstitial atoms, 8–9, 10, 57
Sessile dislocations, 95, 96, 100, 101
Sessile jogs, 149
Shear loops, 146
Shear modulus, 66
Shear strains, 64
Shear stress, 65–6, 82
Shielding dislocations, 246–7

Shockley partials, 85–7, 92, 94, 96
    equilibrium spacing, 154
    face-centered cubic metals, 104
    superlattices, 129, 130, 132
Shuffle set dislocations, 132
Silcox–Hirsch mechanism, 104, 105
Simple cubic structures, 4–5, 15
Simulation methods, 21–2, 34–40,
    199
Single surface helicoids, 16
Size effect interactions, 218
Slip
    asymmetry, 119
    concept of, 43–5
    dislocations, 45–6
Slip bands, 44
Slip planes, 43, 44, 46–8, 164
Slip steps, 164
Slip systems, 239–40, 244
Sodium chloride structure, 10
Solubility limit, 232
Solute atmospheres, 220–8
Solute segregation, 221
Solution strengthening, 233–5
Solution treatment, 232
Space lattice, 1, 2
Spiral dislocations, 57–9
Stacking-fault nodes, 153–4
Stacking-fault tetrahedron, 103–6
Stacking faults, 10–11, 30, 90,
    109–12
    energy, 11, 88
    Frank partial dislocations, 94–9
    shrinkage of, 57
Stacking sequence, 4, 5–6
Stair-rod dislocations, 102–3
Stored energy, 171
Strain aging, 222
Strain energy, 67, 71–3, 184–6
Strain hardening, 99, 205, 237–43
Strain rate, 207–10
Strength of crystalline solids,
    205–49
Stress, definition, 65
Stress concentration, 160–3
Stress fields, 67–70, 174, 176–7,
    181–4
Stress intensity factor, 246
Stress–strain curves, 205–7, 222,
    225–6, 228, 238, 243
Strong forces, 229, 232
Substitutional defects, 9, 10
Superjogs, 142–4, 146
Superlattices, 126–30

Superpartial dislocations, 128–30
Supersaturation, 159, 232
Surface methods, 28–30
Suzuki effect, 220
Symmetrical tilt boundaries, 174, 175, 194–5

T
Tangles of dislocations, 171
Taylor factor, 244
TEM *see* Transmission electron microscopy
Temperature, flow stress, 207–10
Tensile tests, 224–5
Thermal activation, 209–10
Thermal etching, 29
Thermal jogs, 55
Thin foil samples, 28
Thompson tetrahedron, 93–4, 103, 105
Tilt boundaries, 174, 175, 177, 190, 194–5
Transformations, 201–2
Transmission electron microscopy (TEM), 22–8, 198
  dislocations, 24–5

image simulation, 26–8, 29
lattice imaging, 26
limiting factors, 28
planar defects, 25–6, 27, 28
principles, 22–4
Triangular Frank loops, 106
Twin boundaries, 12–13, 20
Twinning dislocations, 115, 122, 195–9
Twist boundaries, 176, 177, 180
Two-beam approximations, 23

U
Unfaulting reactions, 98
Unit cells, 1, 2, 4, 5, 6
Unit dislocations, 18, 86
Unit edge dislocations, 86
Unit jogs, 125, 139–42

V
Vacancies, 8–9, 10
  energy, 80
  jogs, 140
  loops, 29, 115–18, 145
  supersaturation, 159
Velocity of dislocations, 50–3, 210

Visualization, 35
Volterra dislocations, 45
Volume defects, 13
Von Mises condition, 244

W
Wall dislocations, 183–4
Weak beam technique, 25, 27
Weak forces, 229, 231
Whiskers, 14
Work hardening, 99, 205, 237–43

X
X-ray diffraction topography, 31–2, 35

Y
Yield, 210, 220–8
Yield drop, 224–7, 242–3
Yield stress, 127, 205, 224, 234
Yield stress anomaly, 127
Young's modulus, 66

Z
Zonal twinning dislocations, 115

Printed in the United States
By Bookmasters